Xanthates and Related Compounds

Xanthates and Related Compounds

S. RAMACHANDRA RAO
INDIAN INSTITUTE OF TECHNOLOGY, DELHI
NEW DELHI, INDIA

MARCEL DEKKER, INC., New York 1971

To

Dr. C. C. PATEL

of the Indian Institute of Science, Bangalore, who introduced me to Xanthates, initiated me into their investigation, inspired me to their deeper understanding, and helped and guided me in ever so many ways

MARCEL DEKKER, INC.,
95 Madison Avenue, New York, New York 10016

Library of Congress Catalog Card Number 77-141626

ISBN 0-8247-1563-2

PRINTED IN THE UNITED STATES OF AMERICA

PREFACE

Although xanthates were first synthesized in 1822 and many of their chemical properties were known in the last century, the chemical industry had to wait until the end of the century to recognize their importance. The entry of xanthates into chemical technological processes took place in 1892 when xanthates of cellulose were introduced in the manufacture of rayon. The next important milestone occurred in 1925 with the introduction of xanthates as mineral collectors in the froth flotation process of concentrating low-grade ores.

With these two major developments in their history, xanthates came to be recognized as one of the major classes of compounds of significant practical interest. This in turn has given rise to intensive physico-chemical studies of these substances during the last four decades.

The pace of research on xanthates and the related compounds, di-xanthogens, has been particularly fast during the last ten years with the availability of modern instrumental methods of analysis. Both fundamental studies as well as extension of their applications and some new applications have seen considerable advancements.

As a consequence of their nature and their applications in diverse fields of chemistry and technology, literature on xanthates is scattered in journals of such diversified fields as structural chemistry, analytical chemistry, mineral technology and textile technology. While this diversification of interests is to some extent natural, it is needless to emphasize that interaction of ideas and developments in different fields wherever meaningful will significantly help in further advancement of knowledge and its applications to further practical use.

It is with this background that I have written the present monograph. The object in mind has been to gather the material available in diversified literature, organize it systematically, bringing forth links between different pieces and make a critical survey of the present state of knowledge. I have also made an attempt to indicate the frontiers of the present-day research and future potentialities in various fields where xanthates are used.

The subject matter is divided into various chapters covering the different aspects of the chemistry and technological applications of xanthates and dixanthogens. This division is to a great extent a matter of convenience. It is, however, hoped that a study of material in all chapters will be of considerable scientific use to all investigators and technical personnel, even when their primary interest is restricted to any one of the fields covered.

In compiling this volume I have freely drawn material from a number of sources. Reproduction of material has been acknowledged in the text. I would like to mention a word of thanks to various organizations and publishers who have readily and freely granted their permission to reproduce such copyright material.

I am highly thankful to a number of individuals who have taken part in this publication programme. In particular I would like to mention Mr. S. S. Saini who did the painstaking job of typing the manuscript and Mr. I. S. Chopra who drew all the figures with meticulous care. The exacting job of preparing the final manuscript for photocomposition was successfully carried out by Mr. Ravinder Singh. Last, but not least, I am deeply grateful to the library staff of Indian Institute of Technology, Delhi for their unfailing help in locating literature and to the authorities of the Institute for providing all facilities while preparing this monograph.

Indian Institute of
Technology, Delhi
November 14, 1970.

S. R. Rao

CONTENTS

CHAPTER I

INTRODUCTION

Xanthates are salts of xanthic acid, which is essentially alkyl dithiocarbonic acid with two oxygen atoms in carbonic acid replaced by sulfur and one hydrogen atom replaced by an alkyl group:

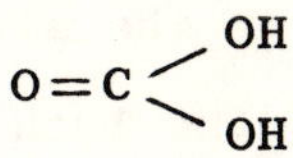

Carbonic acid

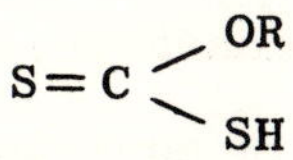

Xanthic acid, (where R denotes an alkyl or aryl group)

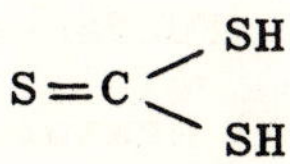

Trithiocarbonic acid

Based on its structure, xanthic acid is also called alkyl dithiocarbonic acid and the corresponding salts alkyl dithiocarbonates. In continental literature xanthates are also referred to as xanthogenates, which was in fact the older name found in 19th century literature.

The hydrogen atom of xanthic acid is replaceable by a metal atom. Furthermore, it can also be replaced by a second alkyl group when a xanthic acid ester or a dialkyl xanthate is derived. Metallic xanthates are most well known as reagents in mineral flotation and have also been extensively used in analytical chemistry. Alkali and alkaline earth metallic xanthates are fairly soluble in water and are strong electrolytes. Xanthates of transition metals are sparingly soluble. Cellulose xanthates find extensive application in the textile industry in the production of viscose rayon.

Xanthates were discovered by Zeise[1] in 1822, but not much work was recorded on the study of their properties and reactions during the 19th century, although some of their general characteristics such as instability in air and in aqueous solution had presumably been recognized. Fundamental studies involving physicochemical measurements on xanthates were begun in the early years of the present century. Work on the stability of xanthic acids was systematically reported in 1913, and the dissociation constants of various xanthic acids have since been determined. The chemical reactions of xanthates and the kinetics of these reactions were studied at different periods between 1920 and 1960. The introduction of xanthates as mineral collectors in the 1920s provided a strong impetus to the detailed study of these compounds. Methods of preparation, purification, and analysis were standardized for the fundamental studies. Theories of the collector action of xanthates (discussed in Chapter 7) have also served to intensify surface chemistry studies on xanthates, along with the determination of physicochemical constants such as solubilities, rates of oxidation and interaction with metal ions under different conditions. Investigations of the collector action of xanthates are largely concerned with the nature of the adsorbed film, relative solubilities of heavy metal xanthates, and the solution chemistry of xanthates under the complex conditions of flotation. Significant progress in such physicochemical studies has been made during the last 10 years through the application of modern instrumental techniques such as electron diffraction, infrared spectroscopy, and X-ray diffraction. These methods have also aided in the elucidation of the structure of xanthates in the crystalline and adsorbed states. Work in this direction is still in progress.

Although many of the studies on xanthates have been carried out with a view toward explaining their action in mineral flotation, xanthates have also attracted the attention of analytical chemists who have made use of their formation in the detection and estimation of alcohols and carbon disulfide. Xanthates have been widely employed as reagents in the separation and quantitative determination of a large number of cations. These procedures are based on the characteristic colors and low solubilities of a large number of heavy metal xanthates and the differential solubilities under controlled pH conditions. At one time xanthates were proposed as reagents in a qualitative analysis technique to be substituted for one based on the use of hydrogen sulfide. The use of metallic xanthates as analytical reagents is still growing, particularly since the development of instrumental colorimetric methods that can detect color changes over a wide range of wavelengths.

Among the derivatives of xanthates, dixanthogens, which are formed by the oxidation of xanthates according to the following equation, are by far the most important:

$$2S=\underset{\substack{|\\ OR}}{C}-S^{-} \qquad S=\underset{\substack{|\\ OR}}{C}-S-S-\underset{\substack{|\\ OR}}{C}=S + 2e^{-} \qquad (1\text{-}1)$$

The role of dixanthogens in conjuction with

xanthates in the flotation of sulfide minerals has attracted considerable attention. In view of this, the formation of dixanthogens under various conditions has been investigated (see Chapter 4).

Another large class of xanthate derivatives are the so-called cellulose xanthates, which are xanthates containing a complex cellulose hydroxy group (in place of an alkyl group) and a sodium or potassium atom. As mentioned above, they are intermediates in the production of rayon. Their reactions in the viscose pulp have been the subject of a large number of investigations in the past 30 years.

Xanthates of a few transition metals have been investigated in recent years with a view to determining their composition and structure. Solubility products have been determined by potentiometric and conductometric methods. Knowledge of the chemistry of transition metal xanthates is also helpful in elucidation of the collector mechanisms of xanthates employed in the flotation of minerals of transition metals such as copper, zinc, cobalt, and nickel. Much work still remains to be done on the stability and chemical reactivity of transition metal xanthates.

1. INTRODUCTION

Current interest in xanthates and the related compounds, dixanthogens, has largely developed among investigators in mineral flotation, analytical chemistry, and structural chemistry, and is reflected in the number of papers (more than 20 a year) that appear in journals of chemistry and mineral technology. Judging from the present indications, it is expected that this trend will continue for many more years.

REFERENCE

1. Zeise, W.C: Vorleufige Anzeige einer neuen Klasse Von Schwefelverbindungen. Schweigger J. Chem. Phys. 35, 173 (1822); Die Xanthogensaure nebst einigen Produkten und Verbindungen derselben. J. Chem. Phys., 36, 1 (1822).

CHAPTER 2

METHODS OF PREPARATION AND PURIFICATION

Xanthates are the reaction products of carbon disulfide, an alcohol, and an alkali. They are formed by combination of the three reactants in the stoichimetric ratio 1:1:1, with elimination of water. The general reaction is:

$$ROH + CS_2 + MOH \longrightarrow H_2O + RO\underset{\underset{S}{\|}}{C}-SM \qquad (2\text{-}1)$$

where R stands for an alkyl hydrocarbon radical and M denotes a monovalent metal such as sodium or potassium.

The details of preparing potassium ethyl xanthate were described by Foster[1], and these methods are generally followed in the laboratory preparation of sodium and potassium xanthates of primary and secondary alcohols. A specified amount of alcohol is measured into a wide mouth conical flask. A slight excess of potassium hydroxide, calculated on the basis of Eq. (2-1), is introduced into the flask in the form of a saturated solution. The flask is cooled in an ice bath to maintain the temperature below 20 °C. This is necessary because the reaction with carbon disulfide is highly exothermic and xanthates are susceptible to thermal decomposition. An excess of carbon disulfide is added gradually while the mixture is simultaneously stirred by a mechanical stirrer. The stirring should be continued for about 15 min after the addition of the required amount of carbon disulfide, which may be added in slight excess to keep the reaction mass in a fluid state. Under these conditions a thick reddish brown mass is formed. The reddish color results from the formation of alkali polysulfides by a side reaction between alkali metal hydroxide and carbon disulfide. The solid is separated by filtration. The reddish orange liquid, consisting of polysulfide impurities, is removed from the solid as much as possible by gentle suction and by simultaneously pressing the solid gently with a porcelain spatula. The solid (which still appears reddish brown) is then dissolved in minimum amount of pure acetone. The polysulfides form a separate red layer. The xanthate also dissolves, giving a pure yellow solution. The two layers, red and yellow, are separated using a separating funnel. To the xanthate solution is added petroleum ether, which acts as a desolvent, and the xanthate is precipitated as a fine faint yellow solid. It is filtered, washed with petroleum ether, and dried in a vacuum desiccator. The process of dissolving in acetone and reprecipitating by petroleum ether may be repeated to improve the purity.

In the above procedure acetone may be replaced by ethyl alcohol and petroleum ether by diethyl ether, which is also a nonpolar solvent. Since alkali metal xanthates are electrolytes, they are soluble in all polar solvents and are precipitated by nonpolar solvents. A nonpolar solvent such as diethyl ether (which is freely miscible in alcohol or acetone) also serves to remove dixanthogen formed by the oxidation of xanthates (see Eq. 1-1). Being nonpolar neutral molecules, the dixanthogens are dissolved in nonpolar solvents.

Gomez[2] has described a method for the preparation of the isoamyl xanthate of potassium which is essentially similar to the one described above. An interesting feature of Gomez's work is his study of the kinetics of the reaction leading to the formation of xanthate. The values of the reaction velocity constants were found to follow a bimolecular reaction rate law and not a trimolecular one, as was expected. This was explained by postulating two steps that add up to the overall equation corresponding to the formation of xanthate:

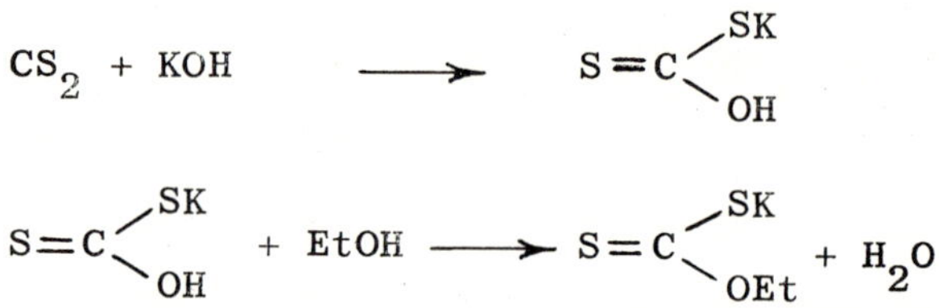

(2-2)

(2-3)

The first of these steps would be expected to be a very fast reaction and would not appreciably influence the overall reaction rate, which is determined by the second step.

As already mentioned, the method of Foster described above is applicable to the preparation of xanthates of primary and secondary alcohols. In the case of tertiary alcohols, however, the method has to be modified because of their lower reactivity. According to Rosenstein[3], a tertiary alcohol such as tributyl carbinol, $(CH_3)_3COH$, is treated with finely ground potassium hydroxide in substantially stiochiometric proportions, a stoichiometric quantity of carbon disulfide is added, and the mixture is heated under a reflux condenser to about the boiling point. As the reaction proceeds, the white alcoholate, $(CH_3)_3COK$, formed by the interaction of alcohol with alkali (see eq. 2-2) changes to the yellow xanthate. The latter is separated by filtration. For the preparation of sodium tert-butyl xanthate, Rosenstein has recommended the preparation of sodium alcoholate by the action of finely divided metallic sodium on tert-butyl alcohol and the addition of carbon disulfide to the latter to obtain sodium tert-butyl xanthate.

A comparative study of the preparation of alkali metal xanthates of various alcohols has been described by Keskyula.[4] In the general method of preparation, 1.2-1.5 parts carbon disulfide are added to1 part potassium hydroxide and 1 part alcohol. The temperature is maintained between 30° and $35^{\circ}C$. The xanthate that separates at once is designated as fraction I the one that separates on cooling as fraction II, and the one obtained by evaporation of the mother liquor as fraction III. The directly separated xanthate (fraction I) is of about the same quality when produced with various kinds of alcohol.

Fraction III is of poor quality (evidently because of contamination by products of thermal decomposition) and its properties depend on the temperature and the degree of evaporation of the mother liquor. It is considered satisfactory for using as a flotation agent (see Chapter 7). The yield of fraction I depends upon the concentration of alcohol, potassium hydroxide, and the water content; the yield is higher the lower the excess of alcohol and the lower the water content. The highest yield (95%) is obtained with a concentration of 1 part potassium hydroxide to 1.5 parts rectified spirit. The total yield of fractions I, II, and III is 95-100%. The mother liquors are unstable, especially when the water content is high, and therefore have to be evaporated immediately after separation of fractions I and II. The reaction could be carried out equally well in glass or iron vessels. Instead of alcoholic potassium hydroxide, aqueous alcoholic potassium hydroxide could be used and the xanthate could be obtained with a 75-78% yield; the purity of the product is between 94 and 99%. With aqueous alcoholic potassium hydroxide, however, the mother liquor is found to be less stable (probably because of hydrolysis). This causes difficulty in the isolation of the xanthate remaining in solution. In general, the yield of fraction I is better the less the excess of the reactants. This fraction is best separated by centrifuging and drying at 50°-60°C. For the production of 1000 kg of xanthate, using ground potassium hydroxide, about 520 kg of potassium hydroxide, 540 kg of ethyl alcohol, and 550-600 kg of carbon disulfide are required, the molar ratio of alcohol to potassium hydroxide being nearly 1:1.

Sodium ethyl xanthate could be prepared as simply as the potassium salt. The reaction is carried out with the smallest possible excess of alcohol and carbon disulfide (nearly 1.1 moles of each per 1 mole of sodium hydroxide). The homogenous solution of sodium hydroxide in alcohol is first prepared, and carbon disulfide added at such a rate that the temperature is kept between 30° and 35°C. After all the carbon disulfide is added, the mixture is stirred until it is solidified. The raw product is dried in thin layers in the air at 15°-20°C for 24 h, and then for 2-3 days at temperatures up to 70°C. In this way 90-95% of a product containing 80-90% sodium xanthate is obtained. In the presence of diluting agents such as benzene or carbon tetrachloride, the reaction occurs even at 10°-$20^\circ C$ and with better yields, but the loss of solvent is considerable.

Keskyula has given the percentage yields of xanthates of a large number of higher alcohols. Xanthates of many alcohols and technical mixtures of alcohols could be prepared under the same conditions as those described for ethyl xanthate by mixing the respective alcohol with the powdered alkali and carbon disulfide. The reaction varies with the different alcohols. Primary alcohols react almost completely, especially those with side chains. Secondary alcohols react less readily, the reaction does not go to completion and by-products are formed, especially if the alcohol contains water or unsaturated compounds (for example, alcohols derived from cracking products). Tertiary alcohols react only at high temperatures (as described by Rosenstein, see above) and yield small amounts of xanthates in addition to thiocarbonates and polysulfides. Alcohols react energetically and more completely with potassium hydroxide than with sodium hydroxide. Potassium xanthates can be dried in 2-3 days at 30–70°C without decomposition, but sodium xanthates strongly absorb the mother liquor and decompose readily when warmed, especially when drying is first begun. They can be dried only by carefully raising the temperature up to 60°C within 7-15 days. Upon further warming they undergo spontaneous thermal decomposition. The addition of excess alcohol or water facilitates the reaction and gives a more uniform product, but the addition of excess water is not to be recommended because of the possibility of hydrolysis of the xanthate formed. Furthermore, the recovery of excess alcohol from the mother liquor is very difficult, and since the wasted alcohol raises the cost of production excess alcohol is not employed. Usually, therefore, potassium hydroxide, alcohol, and carbon disulfide are used in the molar ratio 1:1:1, with the addition of a further mole of water in the case of less-soluble xanthates (for example, xanthates of higher alcohols).

PURIFICATION OF XANTHATES

Normally the product obtained by one of the methods described above is adequate for technical uses such as in mineral flotation and in fungicides. The percentage purity of such commercial xanthate is 90-96% and it contains an oxidation product (dixanthogen), polysulfides, and other sulfur compounds formed by side reactions in small quantities. Removal of these impurities and attainment of 99-100% purity is essential for all fundamental studies and standard work on xanthates. For use as analytical reagents (see Chapter 8) also, xanthates must be purified in the laboratory. A method of purification that produces xanthates of 99-100% purity and virtually free from all side reaction products has been described by Dewitt and Roper[5]. This method is very satisfactory for laboratory purification of alkali metal xanthates.

One hundred grams of impure xanthate are added to 1000 ml of warm acetone (40°C) on a water bath and stirred for a few minutes; some of the solid may not dissolve and this undissolved portion is quite different from the rest; it may be a suspended impurity in the commercial xanthate. It is filtered off through a hot water funnel and discarded.

To the cold filtrate, benzene (4000 ml) is slowly added with stirring, and the precipitate is filtered out on a Büchner funnel with the aid of suction. This precipitate is then dissolved in 500 ml of acetone, and 5000 ml of petroleum ether are added rapidly with no stirring. The precipitate thus obtained is again recovered by filtration through a Büchner funnel and washed with several 50-ml portions of petroleum ether (40°-$50^{o}C$). It is then transferred to a vacuum desiccator containing a small beaker of concentrated sulfuric acid and a small amount (15-20 g) of recently heated active charcoal on a watch glass (persumably to absorb petroleum ether vapors) and is kept under a vacuum of 1 mm of mercury for at least 12 h. At the end of this time, the acid and the charcoal are replaced by fresh portions and the desiccator evacuated again. Xanthate is stored in a vacuum. This purified product should have no odor, is of a light cream (or faint yellowish) color, and dissolves in water, giving a colorless dilute solution. Dewitt and Roper established the purity of xanthate obtained by this method by comparative determinations of sulfur content. The percentage of sulfur found by the Carius method was 39.98, which is in excellent agreement with the calculated value (40.01%). The decomposition point of the purified materialwas $200 \pm 0.3^{o}C$. This was found to be easily reproducible.

Purified xanthates can also be stored under diethyl ether. When required for laboratory use, the quantity needed is taken out and the ether allowed to evaporate (preferably in a vacuum or in a nitrogen atmosphere) before using the product for standard work.

XANTHATES OF GROUP II AND GROUP III METALS

Xanthates of alkaline earth metals such as calcium, strontium, and barium are less soluble than potassium and sodium xanthates, but they are strong electrolytes and can be prepared by a method similar to the one described above for the preparation of sodium and potassium xanthates.

Xanthates of Group III elements thallium, indium, and gallium have been described.[6-9] They are formed as practically insoluble precipitates on mixing potassium xanthate with an equivalent quantity of a nitrate solution of the Group III element. The xanthate of aluminium is unstable and the precipitate formed on mixing potassium xanthate with potassium aluminium sulfate rapidly undergoes hydrolysis.

XANTHATES OF HEAVY METALS

Xanthates of a large number of transition metals such as copper, zinc, silver, and molybdenum are known. Since they are sparingly soluble in water they are obtained by precipitation following the double decomposition of a solution of sodium or potassium xanthate and a soluble salt of the heavy metal. To obtain them in a state of high purity, the precipitation and filtration should be carried out in a nitrogen atmosphere since heavy metal xanthates in the moist state are very susceptible to decomposition in the presence of oxygen and carbon dioxide. The precipitate may be washed with ice cold deoxygenated water. A nonpolar organic solvent such as benzene or petroleum ether may also be used to remove the adhering dixanthogen generally present in the precipitation of cuprous xanthates (see below). The precipitate is finally dried and stored in a vacuum.

Xanthates of copper have attracted considerable attention because of the variable oxidation state of copper. When a soluble xanthate reacts with cupric salt, cupric xanthate of a dark brown color is first formed; but this product is unstable and rapidly changes to cuprous compounds of intense yellow color. Dixanthogen is simultaneously produced.

2. METHODS OF PREPARATION AND PURIFICATION

$$Cu^{2+} + 2\,S{=}\underset{\displaystyle OC_2H_5}{\underset{|}{C}}{-}SK \longrightarrow Cu\left[S{=}\underset{\displaystyle OC_2H_5}{\underset{|}{C}}{-}S\right]_2 + 2K^{+} \qquad (2\text{-}4)$$

Cupric ethyl xanthate

$$2Cu\left[S{=}\underset{\displaystyle OC_2H_5}{\underset{|}{C}}{-}S{-}\right]_2 \longrightarrow 2\left[S{=}\underset{\displaystyle OC_2H_5}{\underset{|}{C}}{-}SCu\right] + S{=}\underset{\displaystyle OC_2H_5}{\underset{|}{C}}{-}S{-}S{-}\underset{\displaystyle OC_2H_5}{\underset{|}{C}}{=}S$$

Cuprous ethyl xanthate — Diethyl dixanthogen

(2-5)

Quite often the product is in a colloidal state and requires centrifugation. The dixanthogen is then extracted with benzene or diethyl ether and the cuprous xanthate obtained in a pure state.

De Plaza[10] has described two forms of cuprous isobutyl xanthate, one red and the other yellow. The yellow form is prepared by treating potassium isobutyl xanthate with cupric sulfate solution and washing the precipitate successively with water, alcohol, and gasoline or carbon tetrachloride. It is slightly soluble in all organic solvents in the cold, and somewhat more soluble in hot solvents, especially chloroform, carbon disulfide, and benzene. On cooling, it separates from solution in the yellow form when cold and in the dark red form when hot, the original color reappearing on cooling. Pyridine transforms the yellow form of cuprous isobutyl xanthate into the red form and gives a red solution from which yellow needles separate after a short time. To prepare the red form in the solid state, 5g of cupric sulfate crystals are treated with an excess of sodium sulfite and 3.7g of potassium isobutyl xanthate in water, and the precipitate is washed with water and dried over sulfuric acid. This red form is amorphous, insoluble in water, and slightly soluble in alcohol (giving a yellow solution), being soon transformed thereby into the yellow form. The same change occurs slowly on contact with ligroin. Acetone, benzene, and carbon disulfide cause an instant change from the red to the yellow form. The transformation also takes place by the action of heat which further decomposes the yellow sample without melting. De Plaza has attributed the existence of the two forms to a structural difference. The red form has been assigned the structure which indicates that the residual valence

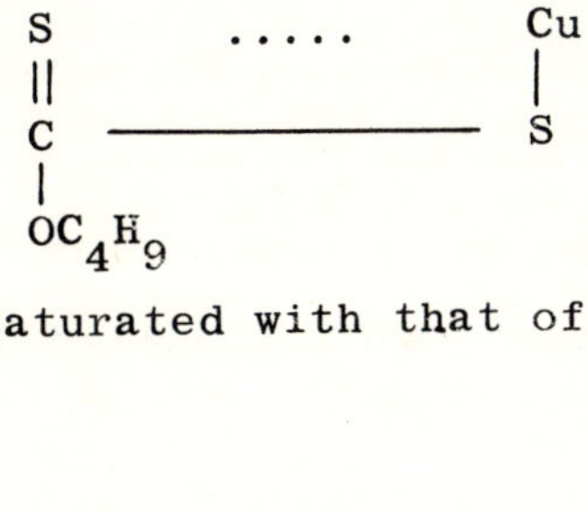

of copper is saturated with that of sulfur. On

contact with organic solvents or by the action of heat, the residual bond (dotted) is ruptured and the red form is transformed into the normal yellow form. Surprisingly, these interesting observations have not received further attention and confirmation. It would probably be worthwhile to determine whether or not the red and yellow forms also occur with other hydrocarbon chains.

Iron would also be expected to form two xanthates, ferrous and ferric, but this has not received serious attention. When potassium xanthate solution is mixed with a soluble ferric salt, a dark brown (almost black) precipitate forms. This is believed to be ferric xanthate although a detailed study has not been described. Ferrous xanthates have comparatively high solubility in water and their preparation in the solid state has not been reported. Presumably they will be oxidized to the ferric state. It would be interesting to determine why a stable ferric xanthate is formed, since under the same conditions cupric xanthate is reduced to the cuprous state and it is well known that the oxidation potential of the $Cu^{+} \rightleftharpoons Cu^{2+} + e^{-}$ system is higher than that of the $Fe^{2+} \rightleftharpoons Fe^{3+} + e^{-}$ system and consequently it would be expected that ferric would be reduced more easily than cupric.

Zinc xanthates also have comparatively higher solubility. Zinc ethyl xanthate is not precipitated in solution but can be prepared by careful crystallization. Zinc amyl xanthate and the higher homologs can be prepared by precipitation. They are white in appearance.

Xanthates of lead occur as yellow precipitates and can be easily filtered on a filter paper. Cobalt, nickel, and other transition metal xanthates are prepared similarly; the precipitates are intensely colored, varying from dark brown to orange yellow.

Silver xanthate is obtained as a yellow precipitate from silver nitrate and potassium xanthate.[11] It is also obtained by leaving silver bromide, silver chloride, or silver iodide in a solution of potassium xanthate for a day.

Arsenic alkyl xanthates are prepared by acidifying with hydrochloric acid a solution of sodium or potassium arsenite at $0^{o}C$. Melting points reported[12] are: isopropyl, $151^{o}C$; propyl, $42^{o}C$; isobutyl, $60^{o}C$; butyl, $38^{o}C$; cyclohexyl, $170^{o}C$ (with decomposition). These compounds are generally very sensitive to the action of alkalis, including ammonia and amines, but are not attacked by acids. They decompose in boiling ethanol, acetone, and ethyl acetate with separation of aresenous sulfide.

Xanthates of noble metals such as platinum and palladium have been described.[13] They are highly insoluble, and this feature is taken advantage of in the separation of the ions of noble metals from those of base metals such as iron.

The syntheses of methyl and ethyl xanthate complexes of platinum (II), palladium(II), nickel(II), chromium(III), and cobalt(III) have been described by Watt and McCormick.[14] The platinum complex is prepared by adding 2 g of potassium chloroplatinate in 50 ml water to 2.35 g of potassium xanthate in 50 ml of water. The bright yellow precipitate is filtered after 3 h, washed with three 50-ml portions of water, and dried over magnesium perchlorate for 2 days. The complex is diamagnetic. The palladium complexes are prepared by a similar procedure, substituting K_2PdCl_4 for K_2PtCl_6; they are also diamagnetic. For the preparation of nickel complexes, 150 ml of methanol containing 4.25 g of nickel acetate tetrahydrate is added to 5 g of potassium xanthate in 170 ml of 53% methanol. The brown precipitate is filtered, washed with three 50-ml portions of water, and dried in vacuo for 2 days. The melting point of nickel ethyl xanthate is 136°-138°C. The product is monomeric and diamagnetic.

Chromic xanthate is prepared by mixing chromium chloride (about 0.03 mole) in 100 ml of water with potassium xanthate (about 0.07 mole) in 30 ml of water. After 2 h the blue precipitate is filtered, dissolved in ethanol, filtered, and evaporated in vacuo to nearly 15 ml. The dark blue crystals are filtered, washed with two 25-ml portions of water, and dried for 2 days in vacuo over magnesium perchlorate. The melting point of ethyl xanthate is 138°-140°C.

Cobaltic xanthates are obtained starting from hexammine cobalt(III) ethyl xanthate, $[Co(NH_3)_6](C_2H_5OS_2)_3 \cdot 3H_2O$. The latter is prepared by slowly adding with stirring 30 ml of a cold aqueous solution of about 12 g of potassium xanthate to 5 g of $Co(NH_3)_6Cl_3$ in 150 ml of cold (3°-4°C) water. After 30 min the precipitate is filtered and washed with 100 ml of cold water, 50 ml of cold ethanol, and 50 ml of cold ether.

A 5.0-g sample of this is warmed slowly to 60°C and held at that temperature for 30 h. The resulting dark green solid is recrystallized once from warm ethanol, washed with small quantities of water, and dried in vacuo for 2 days. The product is monomeric and diamagnetic. The ethyl xanthate of cobalt (III) melts at 123–124°C.

A xanthate complex of molybdenum with the formula $Mo_2O_3(CS_2OC_2H_5)_4$ was described by Malatesta[12] in 1939 and has recently been examined in detail by X-ray diffraction (see Chapter 6). The compound is prepared by mixing 0.2 m ammonium molybdate and 0.4 m potassium ethyl xanthate in 200 ml of water containing about 200 g of ice in the presence of sulfur dioxide gas, which is passed in a slow current. When the precipitation is complete, the gas flow is stopped and the liquid allowed to stand for 4-6 h at a low temperature. It is then filtered under suction, washed with ethanol, and dried over calcium chloride. The solid mass is dissolved in a minimum amount of benzene at 40–50°C and filtered. Double the volume of ligroin is added and the solution is cooled without agitation. The complex separates out as green crystals of melting point 118.5°C (with slight decomposition). It gives an intense violet coloration in various solvents and forms an addition complex with 3 moles of pyridine. The exchange of xanthate groups by cyanide groups on treatment with potassium cyanide has been reported.

The preparation of a potassium uranyl xanthate complex has been described[15]. A solution containing 5.0 g (slightly in excess of 0.03 mole) of potassium ethyl xanthate in 50 ml of water is added to a solution containing 4.2 g (0.01 mole) of uranyl sulfate trihydrate in 150 ml of water with stirring. The uranyl xanthate complex is formed as a brick-red precipitate which should be filtered off rapidly, washed slightly with distilled water, and dried in vacuo over sulfuric acid. For the preparation of isopropyl xanthate complex, a solution containing 0.07 mole of sodium isopropyl xanthate in 75 ml of water is saturated with potassium nitrate. To this, 150 ml 0.1 M uranyl sulfate are added rapidly with stirring. The red precipitate formed is washed and dried as before. The chemical compositions are $KUC_9H_{15}O_5S_5$ and $KUC_{12}H_{21}O_5S_6$, respectively.

The xanthates of transition metals are also characterized by their X-ray diffraction patterns. Tables of X-ray diffraction data are given by Watt and McCormick.[14]

PREPARATION OF XANTHIC ACID ESTERS

The term xanthate ester refers to a double ester with the general formula

$$\mathrm{RO}\overset{\displaystyle\mathrm{S}}{\overset{\|}{\mathrm{C}}}-\mathrm{SR'}$$

where R and R' denote two alkyl radicals which may or may not be identical. They are sometimes referred to as O-alkyl S-alkyl xanthates. The preparation of a large number of compounds belonging to this group has been reviewed by Bulmer and Mann.[16] The general method is to treat sodium or potassium alkyl xanthates with an alkyl halide.

$$\mathrm{S{=}\underset{\displaystyle\mathrm{OR}}{\underset{|}{C}}{-}SK} + \mathrm{R'I} \longrightarrow \mathrm{S{=}\underset{\displaystyle\mathrm{OR}}{\underset{|}{C}}{-}SR'} + \mathrm{KI} \qquad (2\text{-}6)$$

A typical preparation is as follows. Pure alcohol (0.15 mole) is dissolved in 200 ml of reagent grade carbon disulfide contained in a 500-ml three-necked flask fitted with a mercury-sealed stirrer and a reflux condenser. Then, 0.15 mole of pulverized sodium hydroxide is quickly added with continuous stirring and is washed down with 50 ml more of carbon disulfide. The heavy pale yellow gel of sodium xanthate formed is treated with 0.60 mole of methyl iodide in portions, with stirring, and by the next day the gel is transformed into a pale yellow liquid containing a white precipitate of sodium iodide. After dilution with an equal volume of ether and after filtration, the liquid is evaporated under reduced pressure in a vacuum desiccator.

Simple xanthate esters are usually high-boiling liquids, but those derived from aromatic, hydroaromatic, and terpene alcohols are crystalline. The crystalline esters may be purified by recrystallization from a large excess of ethanol and nearly pure white crystals will be obtained.

A series of compounds called sym-dithiocarbonates having the general formula

$$RS-\underset{\underset{O}{\|}}{C}-SR'$$

which are isomeric with xanthic acid esters are known. They are prepared by the action of phosgene on sodium alkyl sulfides or by the acid hydrolysis of alkyl thiocyanates:

2. METHODS OF PREPARATION AND PURIFICATION

$$RS\text{-}Na + R'\text{-}S\text{-}Na + COCl_2 \longrightarrow RSCOSR' + 2NaCl \qquad (2\text{-}7)$$

A dithiocarbonate with identical alkyl groups may also be prepared by the acid hydrolysis of alkyl thiocyanates:

$$2RSCN + 3H_2O \longrightarrow (RS)_2CO + CO_2 + 2NH_3 \qquad (2\text{-}8)$$

REFERENCES

1. Foster, L.S: The preparation of xanthates and other organic thiocarbonates. Utah Eng. Exptl. Sta. Tech. Paper No.2 (1929).

2. Gomez, I: Isoamyl xanthates. Anales Soc. Espan. Fis. Quim. 14, 91-102 (1916).

3. Rosenstein L.: Xanthates and tertiary xanthates. Canadian Pat. 311,334 and 311, 335, May 12 (1931).

4. Keskyula, I.Y.: Potassium and sodium xanthates. Trans. State Inst. Appl. Chem. (USSR), 30, 29-68 (1936).

5. Dewitt, C.C. and E.E. Roper: The surface relations of potassium ethyl xanthate and pine oil. J. Am. Chem. Soc., 54, 444-455 (1932).

6. Anderson, J.R.A.: The preparation and properties of thallous alkyl xanthate. Roy. Australian Chem. Inst. J. Proc. 17, 249-253 (1950).

7. Sutton, G.J.: Indium alkyl xanthates. Roy. Australian Chem. Inst. J. Proc., 17, 249-253 (1950).

8. Sutton, G.J.: Chemistry of some thallium(III) complexes. Australian J. Sci. Res., A4, 654-658 (1951).

9. Delepine, M.: Action of potassium xanthates upon aluminium, gallium and indium salts. Ann. Chim. (Paris,) [12] 6, 645-647 (1951).

10. De Plaza, R.M.D.: Cuprous isobutyl xanthates. Anales Soc. Espan. Fis, Quim., 15, 54 (1917).

11. Reycheler, A: Silver xanthate. Bull. Soc. Chim. Belges, 37, 165-167 (1928).

12. Malatesta, L.: Xanthates of group VI metals and the electric moments of xanthates. Gazz. Chim. Ital., 69, 408-416, 629-639 (1939); 70, 541-553 (1940).

13. Karasev, K.A. and I.A. Kakovskii: Certain sulphdryl compounds of palladium. Tsvetn. Metal., 31, No.3, 47-54 (1958)

14. Watt, G.W. and B.J. McCormick: The synthesis and characterization of methyl and ethyl xanthato complexes of Pt(II), Pd(II), Ni(II), Cr(III) and Co(III). J. Inorg. Nucl. Chem., 27, 898-900 (1965)

15. Zingaro, R.A.: The reaction of uranyl ion with xanthates and dithiocarbonates. J.Am. Chem. Soc., 78, 3568-3570 (1956).

16. Bulmer, G. and F.G. Mann: The chemistry of xanthic acid derivatives. J. Chem. Soc., 666-686 (1945).

CHAPTER 3

GENERAL CHEMICAL PROPERTIES AND REACTIONS

Alkali metal xanthates are very pale yellow solids when pure and have a very faint odor. Xanthates kept in the atmosphere for an appreciable period, however, acquire a deeper yellow color and a rather unpleasant odor somewhat resembling that of carbon disulfide. This is obviously attributable to the slow decomposition of xanthate in the atmosphere, and the products of decomposition, essentially organosulfur compounds, cause the disagreeable odor even when present in low concentrations. Similar behavior is also noticed with xanthates in aqueous solution. When pure, xanthates have appreciable solubility in water and dissolve, giving a clear solution. On standing in atmosphere the solution becomes turbid and gives out a disagreeable odor as a result of hydrolysis and oxidation of the xanthate.

3. GENERAL CHEMICAL PROPERTIES AND REACTIONS

SOLID STATE DECOMPOSITION OF XANTHATES

The decomposition of solid xanthate is enhanced by the action of heat. Thermal decomposition of xanthates was described by Hebert in 1911,[1] but only a qualitative analysis of the products could be made because of the complicated nature of the reaction involving many decomposition products. Furthermore, the mode of decomposition varies with the oxygen content of the atmosphere in which decomposition takes place. Among the products whose formation has been reported are dixanthogens, alcohols, elemental sulfur, dialkyl xanthates, mercaptans, mercaptides, and metallic sulfides.

A detailed investigation of the thermal decomposition of potassium xanthates by gas chromatrographic study and differential thermal analysis (DTA) has been recently described by Tyden.[2] The products of decomposition were separated on two glass columns, one packed with 15% dinonyl phthalate on 7.15 g of Embacel and the other with 16.5 g of silica gel. Chromatographic analysis on a silica gel column showed the formation of hydrocarbons ethane, ethylene, acetylene, and propane and carbonyl sulfide and carbon disulfide. The quantities formed with various xanthates are shown in Fig. 3-1. The formation of carbon dioxide could not be detected. With the dinonyl phthalate column, carbonyl sulfide, carbon disulfide, mercaptans, and alcohols were detected. In addition, aldehydes, sulfides, and disulfides were also formed;

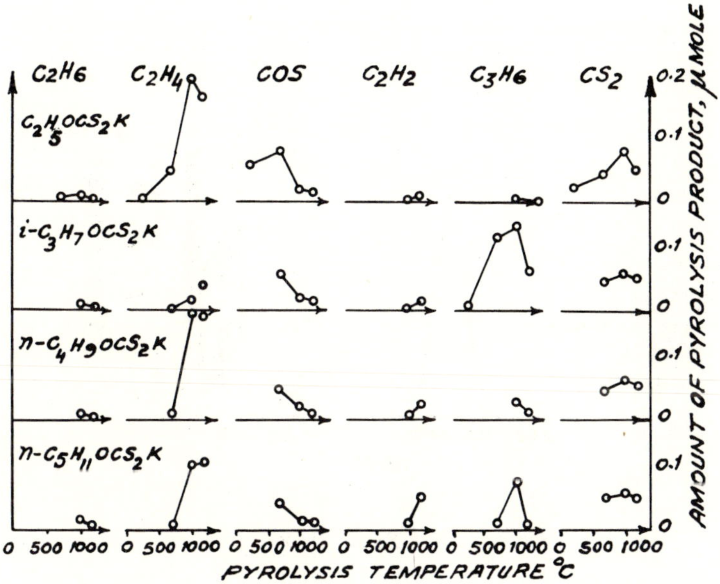

FIG. 3-1

Compounds formed by pyrolysis of potassium xanthates at 225°, 675°, 1000°, and 1225°C. [After I. Tyden: Talenta, 13, 1356 (1966).]

aldehydes could be identified only for n-butyl and sec-butyl xanthates, and for the other xanthates their identification was rendered difficult because of the overlapping carbon disulfide peak. A typical chromatogram is shown in Fig. 3-2, and the quantitative results obtained with dinonyl sulfate are shown in Fig. 3-3. The monosulfides could be positively identified only for methyl, ethyl, and n-butyl xanthate. The identification

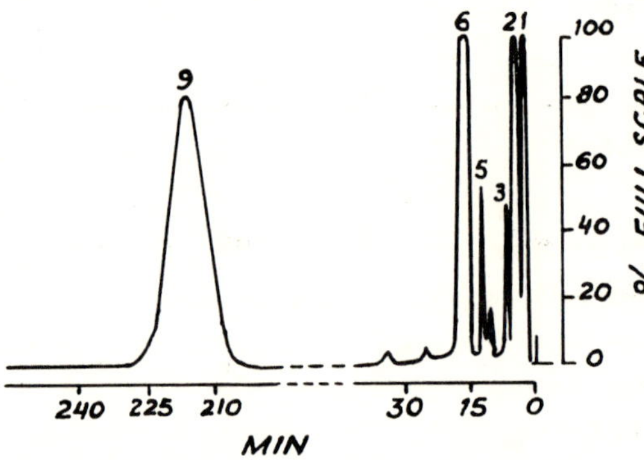

FIG. 3-2

Typical chromatogram obtained from pyrolysis of potassium n-butyl xanthate at 75°C. Peaks: (1) COS + light hydrocarbons, (2) CS_2, (3) n-C_3H_7CHO, (5) n-C_4H_9SH (6) n-C_4H_9OH, (9) $(n\text{-}C_4H_9)_2$ S. [After I. Tyden: Talanta, 13, 1358 (1966).]

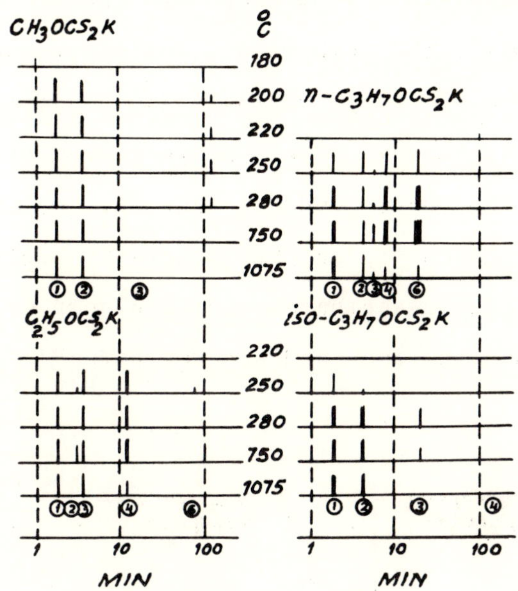

FIG. 3-3

Gas chromatographic separation of pyrolysis products of potassium xanthates. CH_3OCS_2K: (1) COS + light hydrocarbons + CH_3SH, (2) CS_2+ $(CH_3)_2S$, (3) $(CH_3)_2S_2$. $C_2H_5OCS_2K$: (1) COS + light hydrocarbons, (2) C_2H_5SH, (3) CS_2+C_2H_5OH, (4) $(C_2H_5)_2S$, (5) $(C_2H_5)_2S_2$. n-$C_3H_7OCS_2K$: (1) COS + light hydrocarbons, (2) CS_2, (3) n-C_3H_7SH, (4) n-C_3H_7OH, (6) (n-$C_3H_7)_2S$. iso-$C_3H_7OCS_2K$: (1) COS + light hydrocarbons, (2) CS_2 + iso-C_3H_7SH, (3) (iso-$C_3H_7)_2S$, (4) (iso-$C_3H_7)_2S_2$. [After I. Tyden: Talanta, 13, 1358 (1966).]

of the monosulfides from isopropyl and sec-butyl xanthate was rather ambiguous, and the monosulfide from sec-butyl xanthate was produced in too small an amount. The long retention time of these two compounds rendered their detection beyond the reach of the chromatographic column employed. The n-xanthates form monosulfides more readily than the iso- and sec-xanthates probably

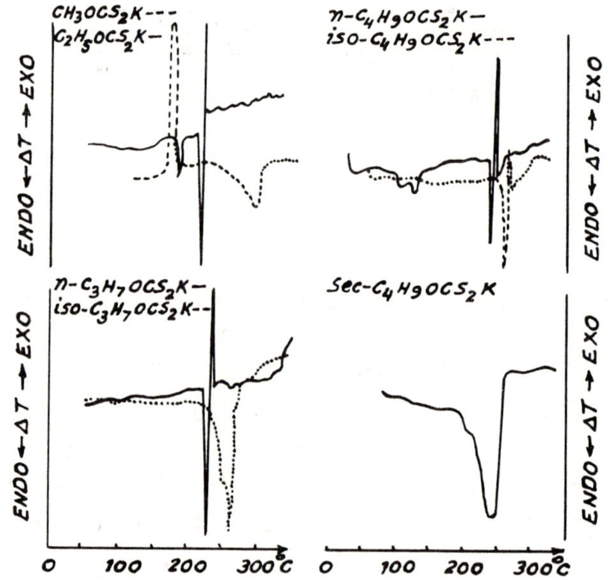

FIG. 3-4

DTA curves of xanthates. [After I. Tyden: Talanta, 13, 1359 (1966).]

because the iso- and sec-hydrocarbon chains decompose more easily to smaller components. The amounts of mercaptan and alcohols as well as the amounts of monosulfides and disulfides decrease with increasing temperature. The decomposition temperatures have been confirmed by DTA. The DTA curves are shown in Fig. 3-4, and the decomposition temperatures of various xanthates are listed in Table 3-1.

TABLE 3-1

Decomposition Temperatures of Potassium Xanthates[a,b]

Hydrocarbon chain of xanthate	T_e, °C	TDTA, °C
Methyl	180-200	165-185
Ethyl	220-250	210-225
n-Propyl	220-250	220-240
Isonpropyl	220-250	230-275
n-butyl	250-280	235-255
Isobutyl	220-250	250-265
sec-Butyl	220-250	220-260

a After I.Tyden: Talanta,13, 1353-1360(1960).

b T_e, Temperature interval in which pyrolysis starts. TDTA, decomposition temperature. The interval is taken from the start to the end of the transition.

3. GENERAL CHEMICAL PROPERTIES AND REACTIONS

DECOMPOSITION OF XANTHATES IN SOLUTIONS

The decomposition of xanthates in aqueous solution is of much greater interest largely because of its importance in mineral flotation (see Chapter 7), and this reaction has been analyzed in recent years from the viewpoint of kinetics as well as thermodynamics. Broadly the decomposition of xanthate in aqueous medium follows three different courses depending upon the pH of the solution and the presence of dissolved oxygen.

On dissolution xanthates dissociate, forming cations of alkali metals and xanthate anions. Although they are strong electrolytes and xanthic acid is a fairly strong acid (as mentioned later), the solution undergoes slow hydrolysis giving rise to xanthic acid which further decomposes into carbon disulfide and alcohol:

$$\text{R-O-C}\begin{smallmatrix}\diagup \text{S} \\ \diagdown \text{S}^- \end{smallmatrix} + H_2O \longrightarrow \text{R-O-C}\begin{smallmatrix}\diagup \text{S} \\ \diagdown \text{SH} \end{smallmatrix} + OH^- \qquad (3\text{-}1)$$

$$\text{R-O-C}\begin{smallmatrix}\diagup \text{S} \\ \diagdown \text{SH} \end{smallmatrix} \longrightarrow CS_2 + ROH \qquad (3\text{-}2)$$

As could be easily predicted from Eq. (3-1), the decomposition of xanthate in this manner would be enhanced by lowering the pH of the solution to neutralize the OH^- ions produced. This happens in the presence of carbon dioxide or when the pH of the solution is lowered by acids.

In the presence of oxygen xanthate is oxidized to dixanthogen according to the reaction:

$$2S{=}C\begin{matrix}S^- \\ OR\end{matrix} + \tfrac{1}{2}O_2 + H_2O \longrightarrow \underset{\displaystyle OR}{S{=}C}{-}S{-}S{-}\underset{\displaystyle OR}{C{=}S} + 2OH^- \qquad (3\text{-}3)$$

The resulting dixanthogen causes turbidity in a xanthate solution kept standing in the atmosphere. As discussed later, the extent of this reaction is very small; the equilibrium is reached after about 5-10% of the xanthate is oxidized in a neutral medium. The extent of the reaction is obviously dependent on the pH of the solution, increasing with a fall in the pH of the medium. In alkaline medium above pH 9 the reaction is virtually suppressed.

In neutral and alkaline media, xanthates decompose by another route, which will be referred to as hydrolytic decomposition. This mechanism was first described by Klauditz[3] and is of interest in the application of xanthates in viscose ripening (cf. Chapter 9). In neutral or mildly alkaline solutions, the following reaction occurs:

$$6S{=}C\langle^{S^-}_{OR} + 3H_2O \longrightarrow 6ROH + CO_3^{2-} + 3CS_2 + 2CS_3^{2-} \quad (3\text{-}4)$$

The thiocarbonate may split up into sulfide and carbon disulfide. It is probable that this reaction could follow from reactions (3-1) and (3-2) by the neutralization of carbon disulfide by hydroxyl ions formed, giving rise to carbonate and thiocarbonate ions:

$$3CS_2 + 6OH^- \longrightarrow 2CS_3^{2-} + CO_3^{2-} + 3H_2O \quad (3\text{-}5)$$

If Eqs. (3-1) and (3-2) are separately multiplied by 6 and the resultant equations added to Eq. (3-5), Eq. (3-4) will be obtained. Since hydroxyl ions are essential for this reaction, it will not take place in an acidic medium.

In strong alkaline solutions, Klauditz[3] identified the following main reaction:

$$S{=}C\langle^{S^-}_{OR} + 5OH^- \longrightarrow 2S^{2-} + CO_3^{2-} + ROH + 2H_2O \quad (3\text{-}6)$$

Both sulfide and carbonate have been identified in the decomposition products. Increasing the hydroxyl ion concentration facilitates greater neutralization of carbon disulfide and, hence, no free carbon disulfide, appears in the last reaction which can also be expressed in two steps:

$$S{=}C\begin{matrix} S^- \\ OR \end{matrix} + H_2O \longrightarrow CS_2 + ROH + OH^- \qquad (3\text{-}7)$$

combination of Eqs. (3-1) and (3-2)

$$CS_2 + 6OH^- \longrightarrow CO_3^{2-} + 2S^{2-} + 3H_2O \qquad (3\text{-}8)$$

Adding Eqs.(3-7) and (3-8) one obtains Eq. (3-6). Other aspects of this reaction with reference to the ripening of viscose will be discussed in Chapter 9.

KINETIC ASPECTS OF THE DECOMPOSITION OF XANTHATES IN AQUEOUS SOLUTION

The effects of medium pH, concentration of solutions, and temperature on the decomposition of aqueous solutions of ethyl and isoamyl xanthates were first reported by Homylev.[4] His results fit the general formula pH = KTa where T = half-life of decomposition of xanthates in minutes, and K and a = constants for a given temperature and a specific hydrocarbon chain. According to this formula, the decomposition of xanthates in an alkaline medium is very slow, a fact which was confirmed experimentally by studying the decomposition of xanthates in solution at different pH values.

Homylev also showed that the higher the concentration and temperature of the solution the more intense the decomposition of the xanthate. A molar solution of ethyl xanthate (presumably dissolved in water from which oxygen was not excluded) gave an alkaline reaction with phenolphthalein immediately after its preparation; a 0.01M solution showed alkalinity after several days, while a 0.001M solution remained neutral for several weeks.

The rate of decomposition of xanthate is also influenced by the length of the hydrocarbon chain. Schaum and co-workers[5] showed that higher homologs of xanthates are less rapidly decomposed in water than lower ones.

Kinetics of the decomposition of xanthates in sodium hydroxide solutions studied by Wronski[6] have shown that the rate of decomposition of methyl, ethyl, propyl, and butyl xanthates can be expressed as a function of sodium hydroxide concentration by the equation

$$\frac{-dx}{dt} = k_1x + k_2x\,(NaOH)^2$$

The ratio k_1/k_2 is about 2 for methyl xanthate and about 3 for other xanthates. The values of the constants decrease with an increase in the length of the hydrocarbon chain. This effect of hydrocarbon chain length on the rate of decomposition of xanthates found by Schaum, and associates[5] in neutral medium and by Wronski[6] in alkaline medium was confirmed later by Rao and Patel[7].

An extensive study of the kinetics of decomposition of ethyl, isopropyl, n-butyl, and isoamyl xanthates of potassium in aqueous solution in the presence of different gaseous systems such as oxygen, carbon dioxide, and a mixture of oxygen and carbon dioxide was carried out by Rao and Patel[7-10]. All three modes of decomposition [reactions (3-2)-(3-4)] were examined. It was shown that the decomposition is very significant during the first hour and that it decays following first-order rate law. The oxidation of xanthate by oxygen according to Eq. (3-3) is much less significant than hydrolytic decomposition according to Eq. (3-4) in the absence of carbon dioxide. In the presence of carbon dioxide, however, oxidation to dixanthogen is significant, as is the acid decomposition of xanthic acid according to reaction (3-2). Some of the results obtained with different xanthates are shown in Fig. 3-5. It may be observed that while with primary alcohol xanthates (ethyl, n-butyl, and

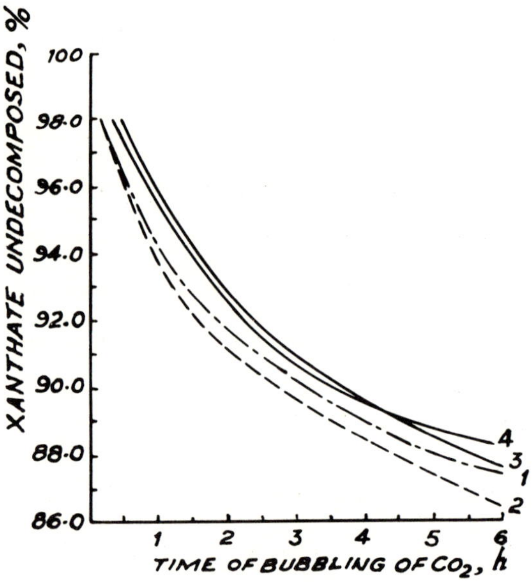

FIG. 3-5

Decomposition of xanthates of different hydrocarbon chains in the presence of carbon dioxide. (1) Ethyl xanthate, (2) isopropyl xanthate, (3) n-butyl xanthate, (5) isoamyl xanthate. [After S.R. Rao and C.C. Patel: J. Mines, Metals Fuels, 8(7), 2 (1960)]

isoamyl) the rate of decomposition decays within 4 h, with isopropyl xanthate prepared from a secondary alcohol the decomposition rate is considerably higher and equilibrium is not reached even after 6 h. The steric effect attributable to the secondary alcoholic group appears to influence the rate of decomposition, although the mechanism has not been satisfactorily explained. The decay in the rate of decomposition of xanthate with time is due to the accumulation of reaction products which inhibit further decomposition. Increasing the pH of the solution tends to diminish the rate of decomposition in the presence of carbon dioxide, but in the absence of it there is appreciable decomposition due to reaction (3-4).

Decomposition of xanthates in aqueous solution is greatly influenced by the presence of metal salts such as those of copper, iron, lead, and zinc which are also commonly present in the flotation pulps of sulfide minerals. The general nature of this effect is shown in Figs. 3-6, 3-7 and 3-8. A discussion of some of the salient features follows.

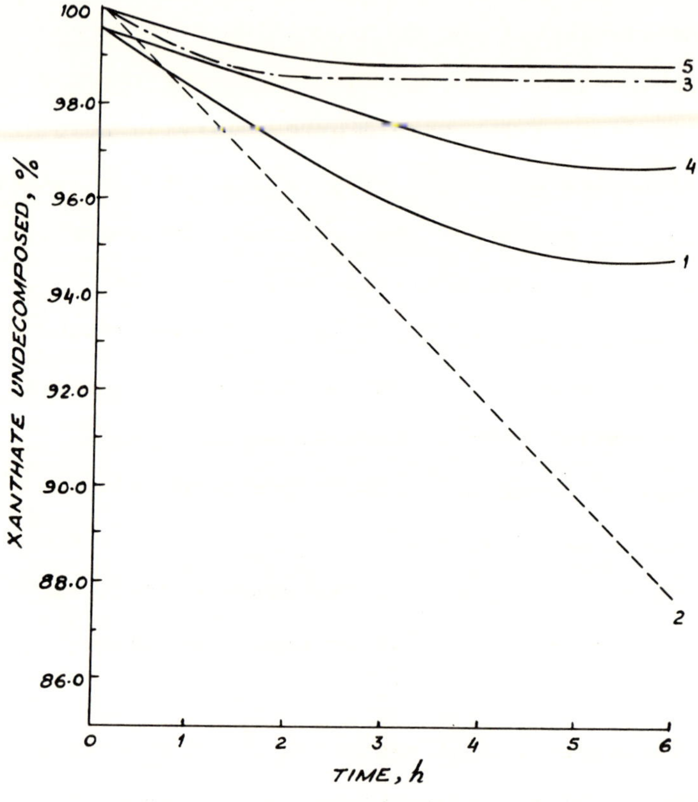

FIG. 3-6

Effect of metal salts on the decomposition of potassium ethyl xanthate by oxygen. (1) In the absence of a metal salt; (2) in the presence of a copper salt; (3) in the presence of a ferric salt; (4) in the presence of a zinc salt.

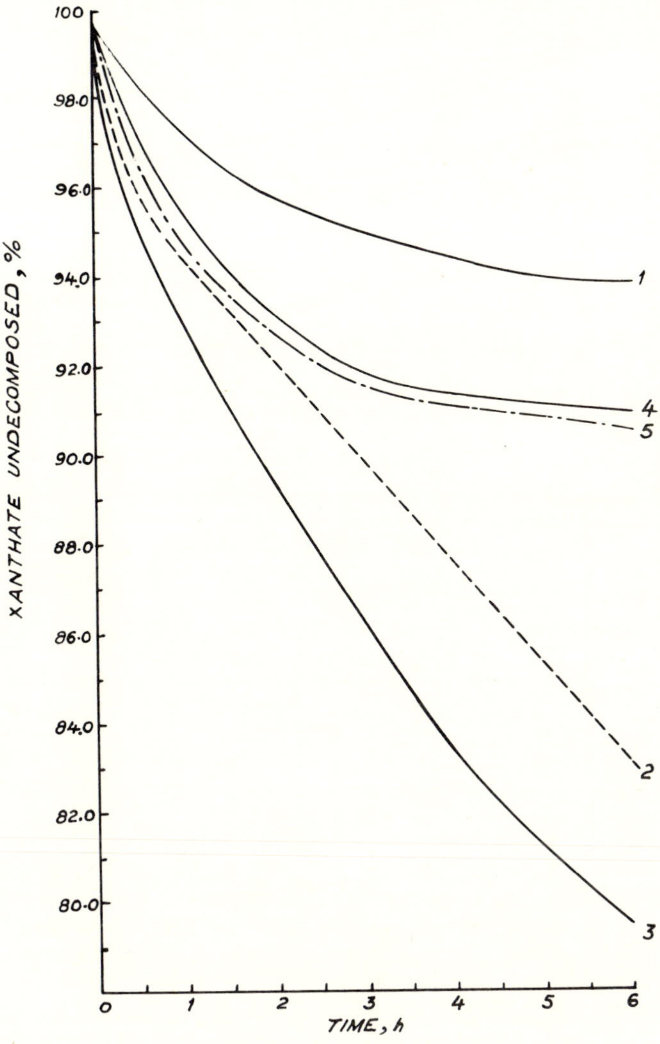

FIG. 3-7

Effect of metal salt on the decomposition of potassium ethyl xanthate by a mixture of oxygen and carbon dioxide. (1) In the absence of a metal salt; (2) in the presence of a copper salt; (3) in the presence of a ferric salt; (4) in the presence of a lead salt; (5) in the presence of a zinc salt.

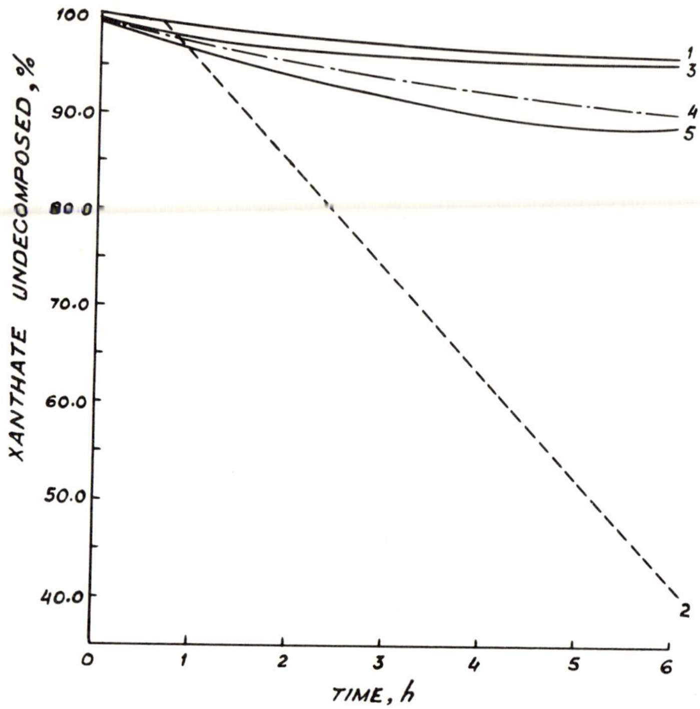

FIG. 3-8

Effect of metal salts on the decomposition of potassium amyl xanthate by oxygen. (1) In the absence of a metal salt; (2) in the presence of a copper salt; (3) in the presence of a ferric salt; (4) in the presence of a lead salt; (5) in the presence of a zinc salt.

CATALYTIC EFFECT OF CUPRIC SALT

It has been observed that in the presence of oxygen the extent of oxidation of xanthates is considerably enhanced with time. This is observed with all xanthates, but with an increase in the hydrocarbon chain of the xanthate, the catalytic effect becomes more marked. This is in contrast to the trend (oxidation) observed in the absence of any salt, when it is found that the extent of oxidation decreases with increase in the hydrocarbon chain of xanthates of primary alcohols. Isopropyl xanthate shows the same type of anomaly in oxidation as was observed in the absence of metal salts. This xanthate is oxidized at a much faster rate than ethyl and butyl xanthates.

The catalytic effect of a copper salt is explained by the following mechanism. Cupric sulfate first reacts with potassium xanthate to give cupric xanthate:

(3-9)

Cupric xanthate, being unstable, decomposes into cuprous xanthate and dixanthogen:

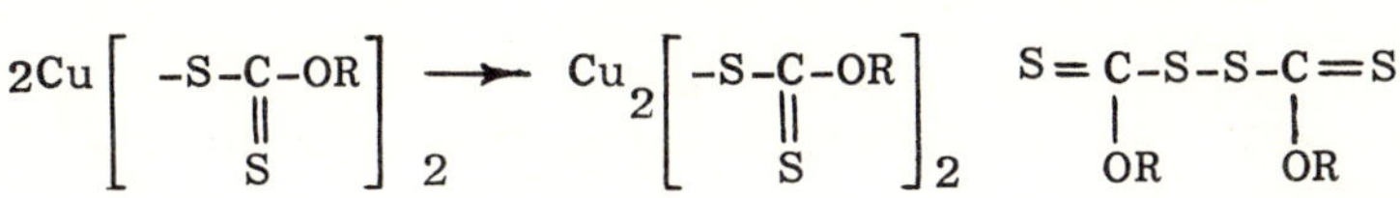

(3-10)

Cuprous xanthate in turn is oxidized to cupric xanthate by oxygen:

$$Cu_2\left[\begin{matrix}-S-C-OR\\ \| \\ S\end{matrix}\right]_2 + \tfrac{1}{2}O_2 + H_2O \rightarrow Cu\left[\begin{matrix}-S-C-OR\\ \| \\ S\end{matrix}\right]_2 + Cu(OH)_2 \qquad (3\text{-}11)$$

Cupric hydroxide, being more soluble than cupric xanthate, reacts with potassium xanthate giving cupric xanthate:

$$Cu(OH)_2 + 2S{=}C\begin{matrix}\diagup SK\\ \diagdown OR\end{matrix} \longrightarrow Cu\left[\begin{matrix}-S-C-OR\\ \| \\ S\end{matrix}\right]_2 + 2KOH \qquad (3\text{-}12)$$

Cupric xanthate again decomposes into cuprous xanthate and dixanthogen, as mentioned above. In this way the whole cycle continues and more xanthate is oxidized in the presence of oxygen.

CATALYTIC EFFECT OF A FERRIC SALT

In the presence of oxygen alone, the catalytic effect of a ferric salt is not as marked. This is attributed to the precipitation of ferric hydroxide (because of the rise in pH of the suspension). Ferric ion concentration is considerably diminished and this reduces the catalytic activity attributable to ferric ions. In the presence of carbon dioxide, ferric ions are available in the solution as a result of the fall in pH which prevents the precipitation of ferric hydroxide. This causes a marked enhancement of catalytic activity. The mechanism put forward to explain the catalytic effect of a ferric salt is as follows. Ferric xanthate is first precipitated by the action of ferric ions on potassium xanthate.

$$Fe^{3+} + 6S{=}C\begin{matrix}\diagup OR\\ \diagdown SK\end{matrix} \longrightarrow 6K^{+} + 2Fe\left[\begin{matrix}-S-C-OR\\ \Vert\\ S\end{matrix}\right]_3 \qquad (3\text{-}13)$$

Ferric xanthate is unstable and is consumed by oxygen and carbon dioxide, giving dixanthogen and releasing ferric ions into the solution:

$$2Fe\left[\begin{matrix}-S-C-OR\\ \Vert\\ S\end{matrix}\right]_3 + 3H_2O + 3/2O_2 + 6CO_2 \longrightarrow 3\,\begin{matrix}S{=}C-S-S-C{=}S\\ |\qquad\qquad\quad |\\ OR\qquad\quad\; OR\end{matrix} + 2Fe^{3+} + 6HCO_3^{-} \qquad (3\text{-}14)$$

The ferric ions readily react with xanthate ions, producing ferric xanthate, and the whole cycle is repeated.

Reactions (3-2) and (3-4) are practically unaffected by the presence of these metal salts.

3. GENERAL CHEMICAL PROPERTIES AND REACTIONS

CATALYTIC EFFECT OF LEAD AND ZINC SALTS

Lead and zinc ions also catalyze the oxidation of xanthate in a manner similar to the mechanism proposed for the catalytic effect of ferric salts.

Kinetics of the decomposition of xanthate in aqueous solution in the presence of oxygen was recently reconsidered by Finkelstein[11] by spectrophotometric analysis (see Chapter 4). Decomposition was followed merely in the presence of air by bubbling nitrogen and by bubbling oxygen. At intervals samples of the solutions were analyzed by readings of the absorbance at the wavelengths 2065, 2260, and 3010A°. The first of these wavelengths is characteristic of aqueous carbon disulfide, and the second and third of aqueous solution of the xanthate radical. Under almost all the experimental conditions, the decomposition of xanthate solutions as indicated by the decrease in absorption at 3010 A° shows good first-order kinetics. The results are summarized in Tables 3-2 and 3-3. Table 3-2 shows the half-lives of decomposition and reaction rate constants for solutions in contact with nitrogen. The rates of decomposition of xanthate in solutions containing dissolved oxygen and their comparison with the corresponding rates obtained in the absence of oxygen are recorded in Table 3-3. The differences between the half-lives of decomposition of solutions in equilibrium with nitrogen and oxygen, $t_{\frac{1}{2}}(N_2) - t_{\frac{1}{2}}(O_2)$, and with nitrogen and air, $t_{\frac{1}{2}}(N_2) - t_{\frac{1}{2}}(\text{air})$ are also shown. Activation energies and frequency factors determined from the reaction rate constants at different temperatures by Arrhenius plots are shown in Table 3-4.

TABLE 3-2

Half-Lives and Reaction Constants for the Decomposition of Xanthate Solutions of Different pH Values in Contact with Nitrogen[a]

pH Value	Temperature °C	Half-life, h	Reaction rate constant, h^{-1} X 10
6	25	37.5	18.5
	35	13.5	51.4
	45	5.5	121
7	25	261	2.65
	35	93.5	7.42
	45	32	21.6

Table 3-2 con't.

8	25	5[illegible]0	1.33
	35	187	3.71
	45	69.0	10.0
9	25	570	1.22
	35	196	3.52
	45	63	11.1
10	25	571	1.21
	35	202	3.45
	45	67.5	10.3
11	25	560	1.24
	35	198	3.5
	45	73	9.5

[a] After N.P. Finkelstein: Trans. Inst. Mining Met., 76, C51 (1967); reproduced by permission of the Institute of Mining and Metallurgy, London.

TABLE 3-3

Effect of the Concentration of Dissolved Oxygen on Xanthate Decomposition Half-lives at Different Temperatures and pH Values[a]

pH Value	Temperature, °C	Half-life with nitrogen $t_{\frac{1}{2}}(N_2)$	$t_{\frac{1}{2}}(N_2) - t_{\frac{1}{2}}(Z)$			
			Z=Oxygen		Z = Air	
			h	%	h	%
6	25	37.5	0	0	0	0
	35	13.5	-0.5	-3.7	0.5	-3.7
	45	6.5	0.2	3.5	0.4	7.0
7	25	261	4	1.5	-2.0	-0.8
	35	93.5	2.5	2.7	2.5	2.7
	45	32.0	0.7	2.2	2.5	7.9

Table 3-3 (Con't.)

8	25	520	0	0	0	0
	35	187	0	0	0	0
	45	69.0	1.9	2.7	9.0	13
9	25	571	38	6.6	-	-
	35	196	8	4.1	-	-
	45	63	0.5	0.8	-	--
10	25	671	44	7.5	-	-
	35	202	12	6.0	-	-
	45	67.5	-0.5	-0.5	-	-
11	25	560	-8.0	-0.15	-	-
	35	193	11	5.5	-	-
	45	73	1	1.4	-	-

TABLE 3-4

Values of Activation Energy and Frequency Factor at Various pH Values for Decomposition of Xanthate Solutions in Contact with Different Atmospheres[a]

pH Value	Atmosphere	Activation energy, kcal $mole^{-1}$	Frequency factor, h^{-1}
6	Nitrogen	17.7	1.8×10^{11}
	Air	18.8	1.1×10^{12}
	Oxygen	17.9	1.8×10^{11}
7	Nitrogen	19.3	3×10^{12}
	Air	20.7	4×10^{12}
	Oxygen	20.7	4×10^{12}
8	Nitrogen	19.2	1.6×10^{11}
	Air	19.8	4×10^{11}
	Oxygen	19.8	5×10^{11}
9	Nitrogen	20.4	8×10^{11}
	Air	-	-
	Oxygen	20.1	8×10^{11}
10	Nitrogen	19.2	1.6×10^{11}
	Air	-	--
	Oxygen	19.3	4×10^{11}

Table 3-4 (con't.)

11	Nitrogen	19.2	1.6×10^{11}
	Air	-	-
	Oxygen	19.5	2.5×10^{11}

[a] After N.P. Finkelstein: Trans. Ins. Mining Met., 76, C51 (1967); reproduced by permission of the Institute of Mining and Metallurgy, London.

The data of Finkelstein show that the differences between the rates of decomposition of solutions with different concentrations of oxygen are small. They point to the conclusion that dissolved oxygen has no effect on the stability of solutions of xanthate in the pH range 6-11. Furthermore, the activation energy of the decomposition reaction is approximately 20 kcal $mole^{-1}$ and is independent of the concentration of oxygen in solution. It is thus inferred that no change in the rate-controlling mechanism of decomposition takes place under the influence of increased concentrations of oxygen in the solution.

Finkelstein attempted to identify the species formed as a result of decomposition of the xanthate at a pH of 10 (without oxygen) by examination of the ultraviolet spectrum. It was found to show two peaks, one at 2200 A^{o} and the second at 2860 A^{o}. The former was found to be identical with that of potassium ethyl monothiocarbonate ($O{=}C(SK)(OC_2H_5)$), the formation of which was thus inferred. The absorption peak at 2860 A^{o} could not be ascribed to any known compound. Although it was close to the reported absorption maximum of diethyl dixanthogen,[12] since the occurrence of the decomposition product was not associated with the presence of oxygen or any other oxidant, the formation of dixanthogen could not be easily accepted. These observations show that in the decomposition of xanthate in alkaline pH range (above pH 8) the mode of decomposition is somewhat different from that suggested earlier, and although no definite mechanism has been put forward it is clear that monothicarbonate and an unknown compound cause complications in the measurements. In fact, the formation of monothiocarbonate as an intermediate product in the decomposition of xanthate in alkaline medium had been suggested by Philip and Fichte,[13] who proposed the following decomposition scheme:

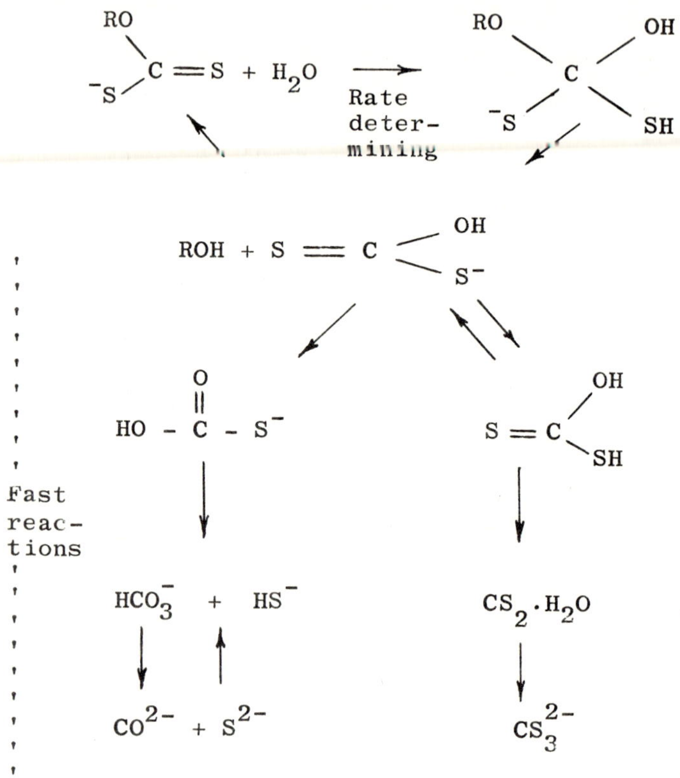

(3-15)

According to Philip and Fichte, the stability of xanthate solutions is dependent on acidity between the pH values of 4 and 7 and almost independent of pH between the values of 7 and 14. A close examination of the half-life periods and reaction rate constants shown in Table 3-2 are nearly in agreement with this although it suggests that the limit of the region in which pH is important is closer to pH 8 rather than pH 7. The dependence of the reaction rate at a lower pH range is attributable to the formation of xanthic acid whose rate of decomposition is increased greatly with a decrease of pH (see next section).

DISSOCIATION AND DECOMPOSITION OF XANTHIC ACIDS

The earliest quantitative studies on the decomposition (Eq. 3-2) and dissociation of xanthic acids were described by von Halban and Kirsch[14] and von Halban and Hecht.[15] The rate of decomposition was found to vary with the nature of the solvent. If the initial rate of decomposition in carbon tetrachloride is taken as unity, the rate at the same temperature is 1.5 in hexane, 3.6 in chloroform, 5 in benzene, 315 in nitrobenzene, about 500 in ethyl ether, 26,000 in acetone, 1 million in ethyl alcohol and over 1 million in water. The reaction is self-accelerating since it is catalyzed by the alcohol produced from xanthic acid. There seems to be no relation between the rate of decomposition and the dielectric constant of the medium although it is observed that the rate increases with the polar character of the solvent. The reaction follows first-order rate law and obeys the Arrhenius equation for the variation of rate constant with temperature.

The rates of decomposition of methyl and ethyl xanthic acids in water at 0°C have been studied.[15] The decomposition was followed by dissolving a known amount of the corresponding sodium xanthate in water and liberating the acid by addition of a slight excess of hydrochloric acid. Then, after a measured interval of time, the solution was neutralized with a cooled solution of sodium bicarbonate and titrated with a standard solution of iodine (cf. Chapter 5). It was found that the addition of sodium, ammonium, and magnesium sulfates to the decomposing xanthic acids in water strongly retards the reaction. The solubility and the dissociation constants of methyl xanthic acid at 0°C are 0.05 and 0.034, respectively. The corresponding values for ethyl xanthic acid are 0.02 and 0.030, respectively.

An extensive study of the kinetics of decomposition of xanthic acids in sodium acetate-acetic acid buffers containing varying amounts of dioxane to vary the dielectric constant were conducted by King and co-workers.[16-18] The reaction may be written:

$$\left(RO\text{-}C\begin{matrix}\nearrow S\\ \searrow S\end{matrix}\right)^- + HAc \longrightarrow ROH + CS_2 + Ac^- \qquad (3\text{-}16)$$

Rate constants were calculated by the equation

$$k'K^o = \frac{2.3}{(b-a)t}\left[(a+c)\log\frac{a}{a-x} - (b+c)\log\frac{b}{b-x}\right]$$

where a, b, and c are the initial concentrations of xanthate, acetic acid, and acetate, respectively, and K_o is the thermodynamic dissociation constant of acetic acid. The formation of a critical complex by a bimolecular reaction of xanthate and H^+ ions has been suggested as the probable rate-determining step; the overall reaction mechanism is expressed as

$$ROCS_2^- + H^+ \longrightarrow X^o \longrightarrow \text{Products} \qquad (3\text{-}17)$$

The variation of the rate constant with the dielectric constant of the medium follows the Christiansen-Scatchard equation

$$\log k' = \log k'_\infty - \frac{Z_A Z_B \epsilon^2}{2.3\ kT\ Dr}$$

where Z_A and Z_B denote the valencies of the cation and anion, D the dielectric constant, ϵ the electronic charge, k the Boltzmann constant,

T, the temperature in ^{o}K, k'_{∞} the rate constant in a hypothetical medium of D = ∞ and r the radius of the critical complex X^{o}. The value of r is calculated from the linear plot of log k' vs. $\frac{1}{D}$.

Reasonable values of r in the range 1.5-2.0 A^{o} were obtained from the experimental data. Values of Gibbs free energy and entropy of activation calculated from the Arrhenius plot for the effect of temperature and the absolute rate equation agree reasonably well with the simple collision theory.

The decomposition of xanthate in acid solution over a wide pH range was reinvestigated by Iwasaki and Cooke[19,20] by using the ultraviolet absorption technique (see Chapter 5) to follow the course of the reaction. Their results, shown in Figs. 3-9 to 3-11, allow the calculation of the dissociation constant of xanthic acid to be made from the spectrophotometric data. The decomposition of xanthate is assumed to take place according to the following steps:

$$H^{+} + X^{-} \underset{k_2}{\overset{k_1}{\rightleftharpoons}} HX \xrightarrow{k_3} ROH + CS_2 \qquad (3\text{-}18)$$

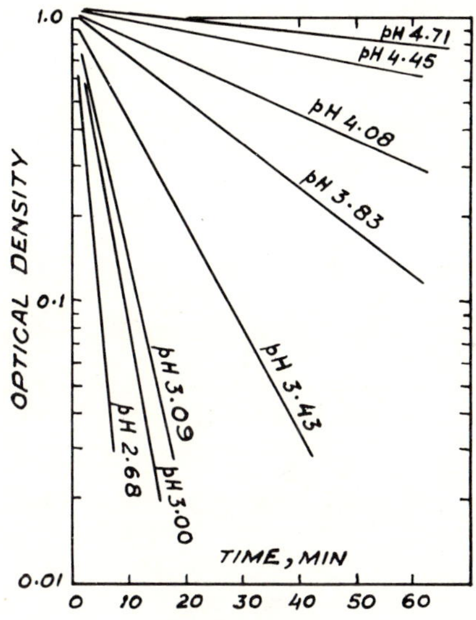

FIG. 3-9

Decomposition of potassium ethyl xanthate in the pH range 2.68 - 4.71. [After I. Iwasaki and S. R. B. Cooke: J. Am. Chem. Soc., 80, 285 (1958); copyright 1958 by the American Chemical Society and reproduced by permission.]

The general rate equations for xanthate, xanthic acid, and carbon disulfide are expressed by the relations

$$\frac{d[X^-]}{dt} = -k_1[H^+][X^-] + k_2[HX] \qquad (3\text{-}19)$$

$$\frac{d[HX]}{dt} = k_1[H^+][X^-] - k_2[HX] - k_3[HX] \qquad (3\text{-}20)$$

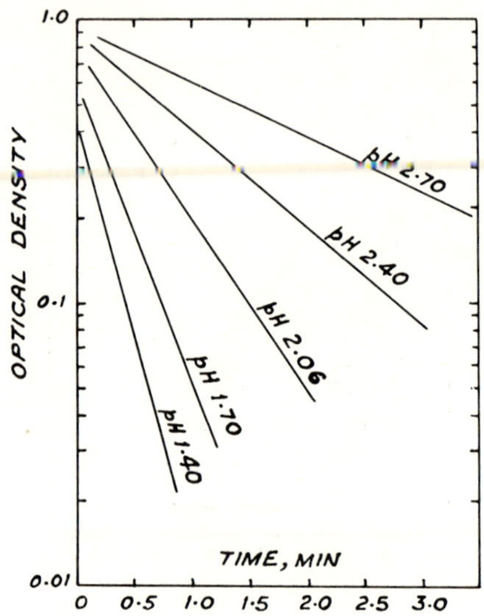

FIG. 3-10

Decomposition of potassium ethyl xanthate in the pH range 1.40-2.70. [After I. Iwasaki and S.R.B. Cooke: J. Am. Chem. Soc., 80, 285 (1958); copyright 1958 by the American Chemical Society and reproduced by permission.]

$$\frac{d[CS_2]}{dt} = k_3 [HX] \tag{3-21}$$

By combining Eqs. (3-19) and (3-20) and expressing the concentration of xanthate ion in terms of total concentration c and the dissociation constant K of xanthic acid, the following expression is deduced:

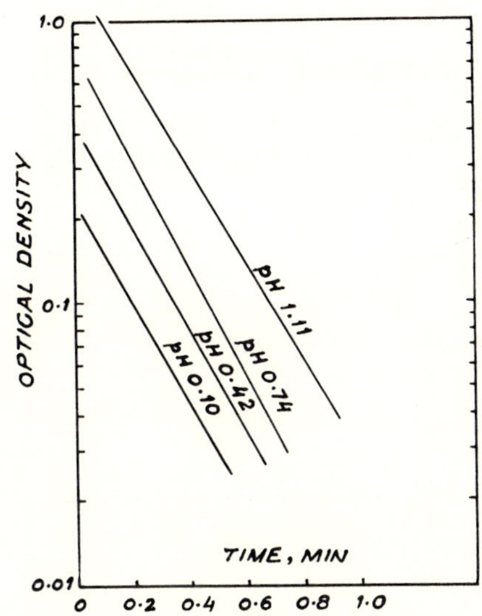

FIG. 3-11

Decomposition of potassium ethyl xanthate in the pH range 0.10-1.11. [After I. Iwasaki and S. R. B. Cooke: J. Am. Chem. Soc., 80, 285 (1958); copyright 1958 by the American Chemical Society and reproduced by permission.]

$$-\frac{d \ln c}{dt} = \frac{k_3}{1 + K/[H^+]} = \varkappa \qquad (3\text{-}22)$$

The values of $\varkappa$ were calculated from the slopes of Figs. 3-9 to 3-11. A plot of log $\varkappa$ vs. pH is found to be a straight line (Fig. 3-12) in the range 2.7-5.0. This agrees with the approximation when

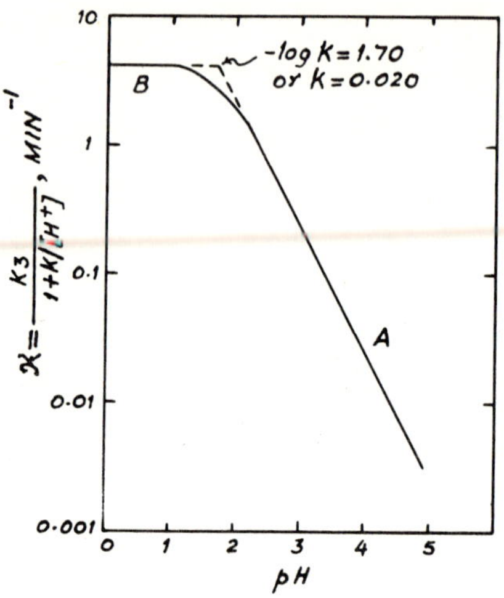

FIG. 3-12

$\mathcal{K}$ as a function of pH

[After I. Iwasaki and S.R.B. Cooke: J. Am. Chem. Soc., 80, 285 (1958); copyright 1958 by the American Chemical Society and reproduced by permission.]

$$[H^+] \ll K, \qquad \mathcal{K} = \frac{k_3}{K} H^+$$

$$\text{or } \log \mathcal{K} = \log \frac{k_3}{K} - \text{pH} \tag{3-23}$$

In the pH range below 2-7, $[H^+] \gg \mathcal{K}$ and hence

$$\mathcal{K} \cong k_3 \tag{3-24}$$

and the average value of $\mathcal{K}$ in the pH range 0-2.7 corresponds to k_3. When this value is substituted in log k_3/K obtained from the linear portion of Fig. 3-12, the dissociation constant K of xanthic acid may be calculated. This value for ethyl xanthic acid is found to be 0.020 ± 0.001, which agrees reasonably well with the values reported earlier.[15,16,20] It should be pointed out that the spectrophotometric method employed by Iwasaki and Cooke[19] is superior to chemical methods of analysis employed by earlier investigators. The chemical methods are subject to errors resulting from the interference of small amounts of other sulfur compounds present as by-products of dissociation.

The reaction mechanism of the decomposition of xanthic acid given by Cooke and Iwasaki implies a unimolecular decomposition of free xanthic acid. A further more detailed analysis carried out by Klein et al.[22] suggests a slightly different mechanism. In their work the concentrations of the xanthate ion and xanthic acid species were determined simultaneously by monitoring their respective absorbances at 302 and 270 mμ. The pseudo-first-order rate constant determined by this analysis shows a maximum at 0.4 N hydrochloric acid. This cannot be explained by the rate equation given earlier, which postulates the decomposition of xanthate via xanthic acid in equilibrium with the xanthate ion (cf. Eq. 3-17). The alternate mechanism proposed is as follows:

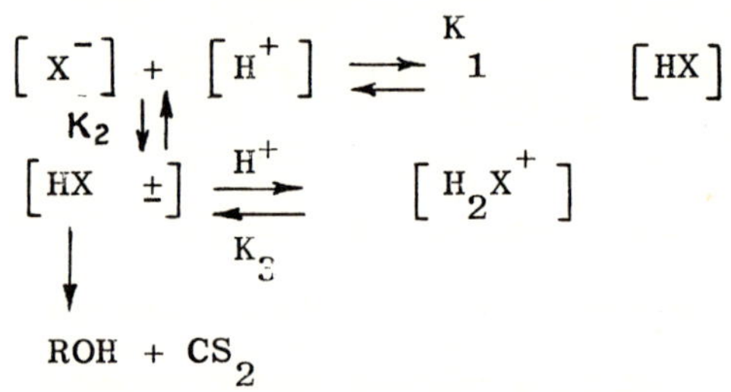

(3-25)

where $[X^-]$ = concentration of xanthate ion, $[HX]$ = concentration of xanthic acid, $[HX\pm]$ = concentration of $C_2H_5\overset{+}{O}CS_2^-$, (ion pair activated complex), $[H_2X]^+$ = concentration of $C_2H_5\underset{H^+}{O}CS_2H$, protonated xanthic acid, and K_1, K_2, and K_3 the equilibrium constants. The formation of protonated xanthic acid is introduced to account for the maximum in the rate curve. The values of rate constants calculated on the basis of this expression agree closely with the experimental values.

From experimental data on the decomposition of xanthic acids at various temperatures, the activation energies have been calculated as a

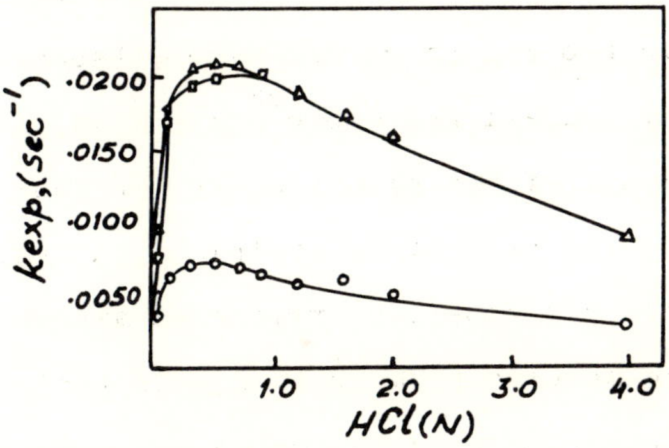

FIG. 3-13

Experimental pseudo-first-order rate constant for the decomposition of ethyl xanthic acid as a function of HCl normality: △, 10°; □, 10°, [Cl⁻] = 0.9 M; ○, 0°. [After E. Klein, J.K. Bosarge, and I. Norman: J. Phys. Chem., 64, 1666, 1960; copyright by the American Chemical Society and reproduced by permission.]

function of acid concentration. The magnitude of activation energy increases with acid concentration; the average value found is 16.9 kcal/mole. The heat of ionization of xanthic acid found is 1.4 kcal/mole. The difference by an order of magnitude between the two values rules out the possibility of identifying the intermediate transient species $HX^{\pm}$ with normal xanthic acid.

3. GENERAL CHEMICAL PROPERTIES AND REACTIONS

The transient formation of protonated xanthic acid and ion pairs has been subsequently supported by Iwasaki and Cooke,[23] who have shown that the values of the dissociation constants of xanthic acid and protonated xanthic acid, and the monomolecular decomposition rate constant of xanthic acid deduced from their spectrophotometric data, are in agreement with the values previously determined.

Törnell[24] has extended the study to a xanthate containing two xanthate groups per molecule such as pentamethylene dixanthate

$$S{=}C(SK){-}O{-}(CH_2)_5{-}O{-}C(SK){=}S$$

A stopped flow technique was used to determine the rates. The dissociation equilibrium was determined by spectrophotometric measurement on acid xanthate solutions in the flowing system. In agreement with Klein and co-workers, the decomposition is considered a bimolecular reaction between dissociated xanthate groups and H^+ ions. The reactivity of a dissociated xanthate group is not markedly affected by the presence of a neighboring undissociated xanthate group, but it increases with the dissociation of this group. This indicates a mutual influence on the reactivity between neighboring dissociated xanthate groups, probably attributable to the influence of the charge of the ion on H^+ ion concentration in the ion atmosphere.

When free xanthic acid is concentrated in nonpolar micelles, the rate of the reaction is decreased because of the exclusion of H^+ ions from the nonpolar micelles. In contrast the presence of surfactant micelles at the constant pH of water increases the reaction rate and the fraction of undissociated xanthate groups. The surfactant micelles in which the xanthate accumulates are regarded as a separate phase characterized by a dielectric constant lower than that in the surrounding water phase and more accessible to H^+ ions, which explains the enhancement of the rate of decomposition of xanthate.

Dissociation constants and decomposition rates for a homologous series of xanthic acids at 25°C and zero ionic strength have been determined by Majima.[25] These are recorded in Table 3-5, which shows that the dissociation constants of the acids and rate constants of the

TABLE 3-5

Dissociation Constants and Decomposition Rates of Xanthic Acids at 25°C and Zero Ionic Strength

Hydrocarbon chain of xanthate	Dissociation constant of xanthic acid, X 10^2	Rate constants for decomposition of acids, litre/mole-min.
Methyl	3.4	233
Ethyl	2.9	226
Propyl	2.5	214
Isopropyl	2.0	207
Butyl	2.3	209
Isobutyl	1.4	202
Amyl	1.9	211

decomposition of ions decrease with increasing hydrocarbon chain length. The iso-compounds have values for these constants, that are smaller than the corresponding normal ones. These values agree reasonably with those determined by Gavrish by employing emf measurements.[26]

Heijl and Pechar[27] have given an emprirical relation between the dissociation constant Kd and the number of CH_2 groups of xanthic acid. This is expressed by the equation,

$$\log (Kd \times 10^4) = a_0 + a_1 X^{-1} \qquad (3\text{-}27)$$

where x denotes the number of CH_2 groups plus 1, and a_0 and a_1 constant coefficients. However, some of the values for dissociation constants reported by Heijl differ markedly from the values reported previously. No explanation has been offered and the empirical relation of Heijl can be taken as a tentative guide in calculating the dissociation constants of xanthic acids that have not been experimentally determined.

REACTIONS OF XANTHATES WITH OTHER REAGENTS

As already mentioned, one of the major characteristics of xanthates is their susceptibility to oxidation, resulting in the formation of dixanthogens. Various mild oxidizing agents such as iodine and hydrogen peroxide are effective in bringing out this reaction. Stronger oxidizing agents such as permanganate are not desirable since they further oxidize sulfur to its higher oxidation state, resulting in the formation of species such as sulfites and sulfates.

The oxidation potential of xanthate increases with an increase in the hydrocarbon chain length. The values obtained by Kakovskii and co-workers[28] are given in Table 3-6.

TABLE 3-6

Oxidation Potentials of a Homologous Series of Xanthates

Hydrocarbon chain of xanthate	E^o, V
Methyl	-0.013
Ethyl	-0.037
Propyl	-0.068
Butyl	-0.100
Amyl	-0.132
Hexyl	-0.157
Heptyl	-0.179
Octyl	-0.203
Nonyl	-0.239
Decyl	-0.276

REACTIONS OF XANTHATES WITH OTHER REAGENTS

The oxidation of xanthate by iodine is widely employed in the routine analysis of xanthates (see Chapter 5). Mellgren[29] has measured the enthalpy of this reaction and compared it with the enthalpy of the well-known reaction between iodine and thiosulfate. The similarity between the two reactions is shown by the following equations:

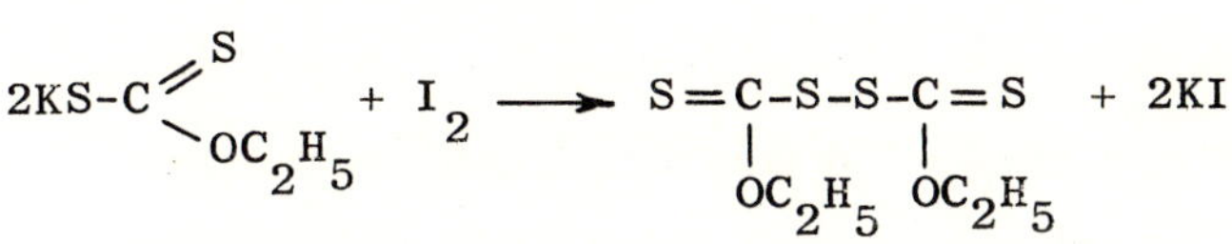

$\Delta H = -18.7$ kcal

(3-28)

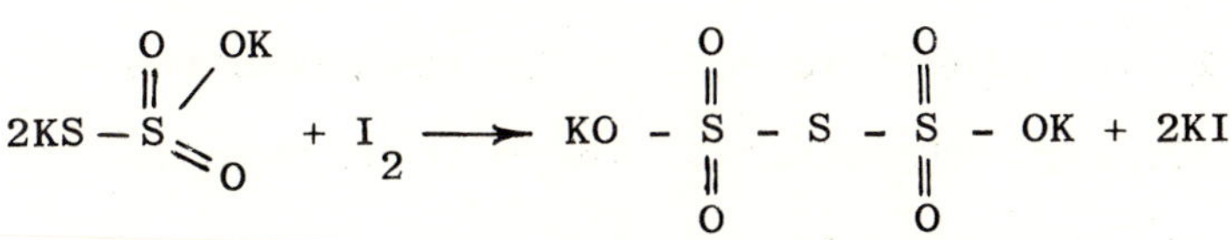

$\Delta H = -7.9$ kcal

(3-29)

An interesting feature of Mellgren's thermochemical studies on xanthate is the finding that ethyl xanthate remains quite stable in the presence of thiosulfate. It was found that a mixture containing 0.0025 m xanthate and 0.005 thiosulfate did not show any change in xanthate concentration when it was kept in air for 21 days. When a mixture of 1 mmole of thiosulfate and 1 mmole of xanthate was titrated with iodine, dixanthogen did not appear, the solution remaining clear until the end point. When thiosulfate was present in slight excess, no turbidity was observed. Titrating a mixture of 3 mmoles of xanthate and 2 mmoles of thiosulfate gave an end point marked by the appearance of dixanthogen after 4 meq of iodine were consumed. A further 1 meq of iodine was consumed. These observations could be explained by postulating the formation of a compound from a xanthate radical and a thiosulfate radical in a 1:1 stoichiometric ratio. It was given the name xanthogen thiosulfate. The heat of reaction leading to the formation of this compound was measured according to Eq. (3-30)

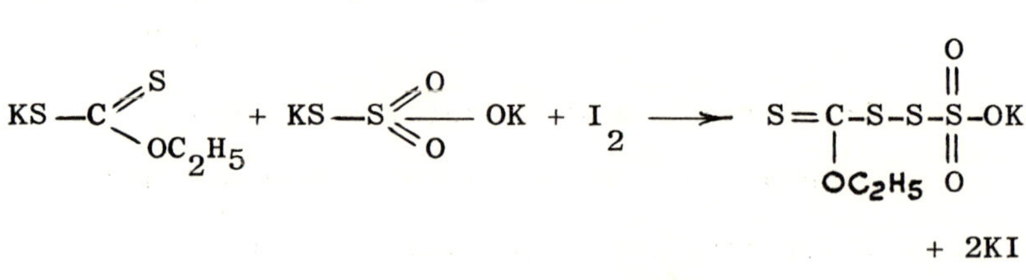

$\Delta H = -14.1$ kcal

(3-30)

The similarity of this reaction to reactions (3-28) and (3-29) is striking. It is also noteworthy that ΔH for reaction (3-30) is only -0.8 kcal larger than the mean value of ΔH for reactions (3-28) and (3-29). It would be interesting to determine whether or not this new compound exists in the free state and whether or not the stability of xanthate in the presence of thiosulfate is related to the formation of a mixed xanthate-thiosulfate ionic species.

Tolum (private communication) has recently pointed out that in the xanthate-iodine-thiosulfate system the following equilibrium is slowly established:

$$2S{=}\underset{\substack{|\\S^-}}{C}{-}OR + S_4O_6^{2-} \longrightarrow 2S_2O_3^{2-} + S{=}\underset{\substack{|\\OR}}{C}{-}S{-}S{-}\underset{\substack{|\\OR}}{C}{=}S \qquad (3\text{-}31)$$

From standard emf data, the equilibrium constant has been estimated to be 2×10^5. The compound "xanthogen thiosulfate" has not been identified by any physicochemical technique. Further work is desirable to clarify the nature of the species present in solution.

In addition to iodine, mild oxidizing agents such as nitrous acid (very dilute), nitrosyl chloride (NOCl), benzene sulfonyl chloride ($C_6H_5SO_2Cl$), tetrathionate, and cyanogen bromide (CNBr) oxidize xanthate to dixanthogen.[30] The action of hydrogen peroxide was studied by Kitamura.[31] In alkaline medium the xanthate is converted into monothiocarbonate by replacement of the double-bonded sulfur atom by oxygen, and the sulfur is oxidized to sulfate at the expense of hydrogen peroxide.

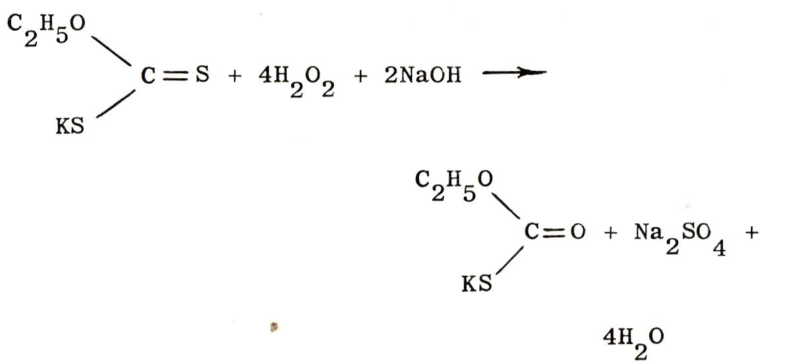

(3-32)

When 10 moles of hydrogen peroxide are mixed 1 mole of xanthate in a 0.01 M solution, the reaction is quantitatively complete. Reaction velocity can be increased by increasing the concentration.

It is advantageous to use twice the calculted amount of alkali.

The action of chloramine-T (the sodium salt of N-chloro-p-toluenesulfonamide) has been studied by Rao and Murty,[32] who have used this reagent for the quantitative determination of carbon disulfide through formation of xanthate (see Chapter 8). In acid medium chloramine-T functions as a strong oxidizing agent since it liberates hypochlorous acid.

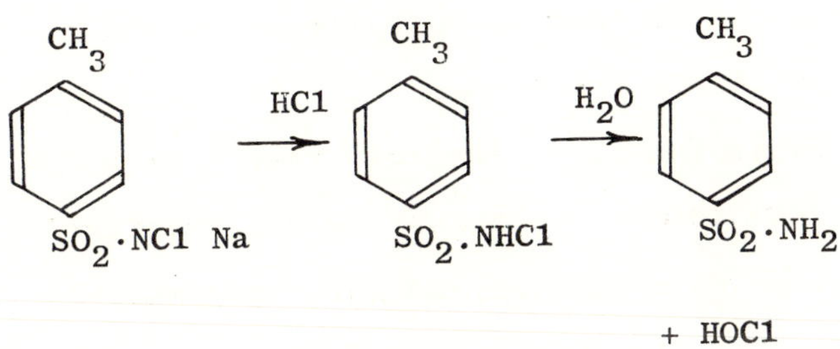

Ethyl xanthate is quantitatively oxidized, 1 mole consuming 14 equivalents of chloramine-T.

$$2S{=}C\begin{matrix} OC_2H_5 \\ SK \end{matrix} + 14\,O + 6H_2O \longrightarrow 2C_2H_5OH + 3H_2SO_4 + K_2SO_4 + 2HCOOH \qquad (3\text{-}33)$$

The reaction with longer-chain xanthates is more complicated because of the steric hindrance attributable to the hydrocarbon chain length.

The formation of a new compound, dinitroso ferrous xanthate, was observed[33-34] when ferrous sulfate in 50% ethyl alcohol was saturated with nitric oxide at $0^{o}C$ and a solution of potassium xanthate added. It has been formulated as $Fe(NO)_2(S{\cdot}CS{\cdot}OC_2H_5)_2$. It is also formed when nitric oxide is passed through an alcoholic solution of ferric xanthate. There appears to be some question as to whether or not iron is in the ferrous or the ferric form.

Potassium ethyl xanthate reacts with benzene-, p-toluene- and naphthalene-2sulfonyl chlorides to form the corresponding potassium sulfonate, the disulfone, and diethyl dixanthogen, respectively.

$$2C_2H_5O\text{-}\underset{\underset{S}{\|}}{C}\text{—}SK + C_6H_5SO_2Cl \longrightarrow S{=}\underset{\underset{OC_2H_5}{|}}{C}\text{-}S\text{-}S\text{-}\underset{\underset{OC_2H_5}{|}}{C}{=}S$$

$$+ C_6H_5SO_2K + KCl \qquad (3\text{-}34)$$

The reaction between potassium ethyl xanthate and p-nitrobenzoyl chloride yields O-ethyl 5-p-nitrobenzoyl xanthate, $C_2H_5O{\cdot}CS{\cdot}S{\cdot}CO{\cdot}C_6H_4NO_2$, which is obtained as a stable crystalline solid.[35]

Unlike xanthates, the xanthic acid esters are often remarkably stable under acidic conditions.[36] They are characterized by their reaction with phenyl hydrazine with which they give O-alkyl-N-phenylamino-thiourethane.

$$ROCSR + C_6H_5NH{\cdot}NH_2 \longrightarrow ROC(=S) - NH{\cdot}NHC_6H_5 + C_2H_5SK \qquad (3\text{-}35)$$

TRANSITION METAL XANTHATES

Literature on the characteristics of transition metal xanthates is rather meager. Preparations have been described in Chapter 2. Xanthates of iron, cobalt, and nickel have been studied by spectrophotometry.[37-39] The ratio Fe^{3+}:xanthate and also that of Co^{3+}:xanthate is shown to be 1:3, and that of Ni^{2+}:xanthate, 1:2. It is somewhat surprising that iron and cobalt are not reduced to the lower oxidation state by xanthate, as copper is known to be reduced to the cuprous state. The average values of the stability constants are 4.36×10^{-9} and 2.08×10^{-4} for iron xanthate and nickel xanthate, respectively.

Many of the transition metal xanthates form complexes with pyridine. Such complexes were described by Dubsky[40] and Delepine[41] nearly 50 years ago. Ferric xanthate forms a complex in which the molar ratio of pyridine to ferric xanthate is 3:1 when dried. The crystals lose pyridine in the air reverting to ferric xanthate. Cobaltic and chromic xanthates are also readily soluble in pyridine, but surprisingly do not form complex crystals with it. Nickelous and zinc xanthates form coordination complexes with 2 moles of pyridine, but these are less stable than those of iron. Cuprous xanthate forms a 1:1 complex. Xanthates of mercury and lead form no complexes. Surprisingly, no further structural studies of such complexes have been undertaken in recent years.

The composition of molybdenum xanthate has been reported by Pilipenko.[42] Trivalent and quadrivalent molybdenum do not react with xanthate. Hexavalent and pentavalent molybdenum react with xanthate in the Mo:xanthate ratios of 1:3 and 1:2, respectively. Since the reaction takes place only in acid solutions in which hexavalent molybdenum is unstable, however, it is inferred that the xanthate compound consists of pentavalent molybdenum only. One molecule of xanthate per 1 molecule of molybdenum is used for the reduction from the hexavalent to the pentavalent state, and the ratio of molybdenum to xanthate remains 1:2.

Uranium xanthate complexes of the formula $KUC_nH_{2n-3}O_5S_6$ with ethyl and isopropyl xanthate groups have been described.[43] These compounds decompose at temperatures above 35°C; they are readily soluble in alcohols, ethers, and ketones, but insoluble in nonpolar solvents. Because of this it has not been possible to establish the structure of the solid compounds from spectral data. The existence of UO_2X_2 in solutions of the solid in ethanol suggests that it may be a double salt of the type $KXUO_2X_2$, in which X is the xanthate group. The spectral data so far obtained however, preclude the possibility of complexes of other compositions, and the structure of the solid is still unknown.

The reaction between xanthate and cadmium ion has been recently studied by Yamasaki and

Nanjo.[44,45] The polarographic studies have revealed the formation of a tetracoordinate complex anion, $[Cd(EtX)_4]^{2-}$ when the precipitate of normal cadmium ethyl xanthate dissolves in excess potassium ethyl xanthate solution. If potassium ethyl xanthate solution is mixed with excess cadium ion, a 1:1 complex, $[Cd(EtX)]^+$ is formed. The stability constant found by spectrophotometric method is 1.8 X 10^4. This complex decomposes in acidic medium:

$$Cd[(EtX)]^+ + H^+ \longrightarrow Cd^{2+} + HEtX \quad (HEtX \downarrow C_2H_5OH + CS_2) \qquad (3\text{-}36)$$

This is the first report on the formation of soluble xanthate complexes. Similar complexes with other transition metals are likely to be of interest in the solution chemistry of xanthates.

Solubility products of a large number of xanthates have been given in terms of activities by Pilipenko[46] and Kakovsky.[47] These are compiled in Tables 3-7 and 3-8.

TABLE 3-7

Solubility Products of Transition Metal Xanthates[a,b]

Compound	Solubility product (in terms of activities of ions), L_0	Activity factor $f = L_0^{1/m}$
AuC_2H_5OCSS	6×10^{-30}	6×10^{-30}
CuC_2H_5OCSS	5.2×10^{-20}	5.2×10^{-20}
$Hg(C_2H_5OCSS)_2$	1.7×10^{-38}	1.3×10^{-19}
AgC_2H_5OCSS	5×10^{-19}	5.0×10^{-19}
$Pb(C_2H_5OCSS)_2$	2.0×10^{-17}	-
$Cd(C_2H_5OCSS)_2$	2.6×10^{-14}	1.6×10^{-7}
$Co(C_2H_5OCSS)_2$	5.4×10^{-13}	7.5×10^{-7}
$Ni(C_2H_5OCSS)_2$	1.4×10^{-12}	1.2×10^{-6}
$Zn(C_2H_5OCSS)_2$	4.9×10^{-9}	7.0×10^{-5}
$Fe(C_2H_5OCSS)_2$	8×10^{-8}	2.8×10^{-5}
$Mn(C_2H_5OCSS)_2$	10^{2}	10^{-1}

[a] From A.T. Pilipenko, and T.P. Varchanko: Zh. Analit. Khim. 12, 457-461 (1957) and A.I. Kakovsky, : In "Proceedings of the Second International Congress on Surface Activity," Vol.4, Butterworth, London, 1957, pp.225-2380. Editor, J.H. Schulman

[b] The activity products of zinc and silver xanthates with chains of different length are listed in Table 3-8.

TABLE 3-8

Activity Products for Xanthates of Different Chain Lengths[a]

Xanthates	Zn	Ag
Ethyl	4.9×10^{-9}	5.0×10^{-30}
Propyl	3.40×10^{-10}	1.4×10^{-19}
Butyl	3.70×10^{-11}	3.6×10^{-20}
Amyl	1.55×10^{-12}	1.5×10^{-20}
Hexyl	1.25×10^{-13}	2.4×10^{-21}
Heptyl	1.35×10^{-14}	-
Octyl	1.50×10^{-16}	1.50×10^{-22}
Isopropyl	2.2×10^{-10}	1.0×10^{-19}
Isobutyl	2.75×10^{-11}	1.6×10^{-20}
Isoamyl	3.10×10^{-12}	5.4×10^{-21}

[a]From A.I. Kakovsky: In "Proceedings of Second International Congress on Surface Activity," Vol. 4, Butterworth, London, 1957, pp. 225-238. Editor, J.H. Schulman

REFERENCES

1. Hebert, A.: Thermal decomposition of metallic xanthates. Bull. Soc.Chim.France, 9, (4) 523-532 (1911).

2. Tyden, I.: Gas chromatographic study of the pyrolysis of potassium salts of xanthic acids. Talanta, 13, 1353-1360 (1966).

3. Klauditz, W.: The ripening of viscose. Papier Fabr. 37, Tech-Wise TI., 251-259 (1939).

4. Homylev, P.I.: The behaviour of aqueous solutions of potassium ethyl- and isoamyl xanthates. Investigatory Work on the Theory and Practice of Flotation, Mekhanobr Inst. (Papers), 2, ONTI (1936).

5. Schaum, K., P.Ziedler and E.Wagner: Hydrolysis and dissociation of xanthates. Kolloid-Z., 65, 264 (1933).

6. Wronski, M.: Kinetics of decomposition of xanthates in sodium hydroxide solutions. Roczniki Chem., 33,1071--1080 (1959).

7. Rao, S.R. and C.C. Patel: Kinetics of decomposition of xanthates. J. Mines, Metals, Fuels, 8(7), 9-14 (1960).

8. Rao, S.R. and C.C. Patel: Kinetics of decomposition of xanthates in presence of ferric salt. J. Mines, Metals, Fuels, 8(7), 15-17 (1960).

9. Rao, S.R. and C.C. Patel: Kinetics of decomposition of xanthates in presence of copper salt. J. Sci. Ind. Res. (India) 20D, 299-303 (1961).

10. Rao, S.R.: Kinetics of decomposition of xanthates in presence of lead salt. J. Sci. Ind. Res. (India), 21D, 125-128 (1962).

11. Finkelstein, N.P.: Kinetic and thermodynamic aspects of the interaction between potassium ethyl xanthate and oxygen in aqueous solution. Trans. Inst. Mining Met., 76C, 51-59 (1967).

12. Pomianowski, A. and J. Leja: Spectrophotometric study of xanthate and dixanthogen solutions. Can. J. Chem., 41, 2219-2230 (1963).

13. Philipp, B. and C. Fichte: Kinetic studies on the decomposition of xanthate. Faserforsch. Textiltech., 11, 118-24, 172-179 (1960).

14. Halban, H. Von and A. Kirsch: Xanthogenic acid and the kinetics of its decomposition. Z. Phys. Chem., 82, 325-360 (1913).

15. Halban, H. von and W. Hecht: Xanthic acids and the kinetics of their decomposition. Z. Elektrochem., 24, 65-82 (1918).

16. King, C.V. and E. Dublon: Rate of decomposition of xanthic acid. J. Am.Chem. Soc., 54, 2177 (1932).

17. Chatenever, A. and C.V. King: Kinetics of decomposition of ethyl xanthic acid. J. Am. Chem. Soc., 71, 3587 (1949).

18. Schochet, M.L. and C.V. King: Kinetics of decomposition of n-propyl xanthic acid. J. Am. Chem. Soc., 77, 4745-4746 (1955).

19. Iwasaki, I. and S.R.B. Cooke: The decomposition of xanthate in acid solution. J. Am. Chem. Soc., 80, 285-288 (1958).

20. Iwasaki, I. and S.R.B. Cooke: Dissociation constant of xanthic acid as determined by spectrophotometric method. J. Phys. Chem., 63, 1321-1322 (1959).

21. Cranendonk, A.C.: Decomposition of potassium ethyl xanthate in mineral acid medium. Rec. Trav. Chim., 70, 431-448 (1951).

22. Klein, E,J.K. Bosarge, and I. Norman: Spectrophotometric determination of fast xanthate decomposition kinetics. J. Phys. Chem., 64, 1666-1670 (1960).

23. Iwasaki, I and S.R.B. Cooke: Decomposition mechanism of xanthate in acid solution as determined by a spectrophotometric method. J. Phys. Chem., 68, 2031-2033 (1964).

24. Törnell, B.: Xanthate decomposition in acid media. Svensk Papperstid., 69, 658, 695 (1966).

25. Majima, H.: Dissociation and decomposition of xanthic acids. Sci. Rept. Res. Inst. Tohoku Univ. Ser.A, 13, 183-197 (1961).

26. Gavrish, A.P.: Dissociation constants of xanthic acids in aqueous solution. Ukr. Khim. Zh., 29,(9), 900-904 (1963).

27. Heijl, V. and F. Pechar: Potentiometric determination of dissociation constants of xanthic acids. Chem. Zvesti, 21(4), 261-266 (1967).

28. Kakovskii, I.A., N.V. Serebrajakova, and B.A. Stepanov: Oxidation reduction potentials of xanthates. Nauchn. Dokl. Vysshei Shkoly, Khim. i Khim. Technol., No.2, 277-279 (1959).

29. Mellgren, O.: Thermochemistry of ethyl xanthate and of its reaction with lead salts. Trans. AIME, 235, 46-53 (1966).

30. Cambron, A. and G.S. Whitby: Oxidation of xanthates. Can. J. Res., 2, 144 (1930).

31. Kitamura, R.H.: Action of hydrogen peroxide on xanthate. J. Pharm. Soc. Japan, 54, 1 (1936).

32. Rao, V.R.S. and A.R.V. Murthy: Determination of carbon disulfide by chloramine-T. Talanta, 4, 206-209 (1960)

33. Manchot, W. and S. Davidson: Nitroso xanthates. Chem. Ber. 62B, 681 (1929).

34. Cambi, L. and L. Szego: Nitroso xanthates, Atti Accad. Lincei, 13, 93 (1931).

35. Bulmer, G. and F.G. Mann: The chemistry of xanthic acid derivatives. J. Chem. Soc., 677-686 (1945).

36. Sanyal, A.K. and C.B. Purves: Some xanthate methyl esters of glucose. Can.J. Chem., 34, 426-435 (1956).

37. Isvoranu-Panait, C.: Spectrophotometric studies of the Fe(III)-xanthate system. Analele Univ. C.I. Parhon Ser.Stiint. Nat., 13 (1), 113-140 (1964).

38. Sheka, Z.A. and E.E. Kriss: The formation of cobalt xanthate. Zh. Neorg. Khim., 1, 586-597 (1956).

39. Korenman, I.M., F.R. Sheyanova, and S.A. Ryabova: Determination of the composition of cobalt xanthate. Zh. Neorg Khim., 2, 65-67 (1957).

40. Dubsky, J.V.: The saturation of the affinities of the principal and secondary valences in compounds of higher order. J. Prakt. Chem. 90, 61-118 (1914); 93, 142-161 (1961); 103, 109-128 (1921).

41. Delepine, M. and L. Conpin: On the xanthates of cobalt and nickel. Bull. Soc. Chim. France, 27, 469-474 (1920).

42. Pilipenko, A.T. and G.I. Gridschina: Composition of molybdenum xanthate. Tr. Komis. P.O. Analit. Khim. Akad. Nauk SSSR Inst. Geokhim. i Analit. Khim., 3, 178-187 (1951).

43. Zingaro, R.A.: The reaction of uranyl ion with xanthates. J.Am.Chem. Soc., 78, 3568-3570 (1956).

44. Yamasaki, T. and M. Nanjo: A polarographic study of the reaction between a xanthate and cadmium ion. Chem. Ind. (London), 1530-1531 (1966).

45. Nanjo, M. and T. Yamasaki: Spectroscopic and kinetic studies of reactions of potassium ethyl xanthate with cadmium ion. Chem. Ind. (London), 1454-1455 (1967).

46. Pilipenko, A.T. and T.P. Varchanko: Solubility product of zinc, nickel, iron and cadmium xanthogenates. Zh. Analit. Khim., 12, 457-461 (1957).

47. Kakovsky, I.A.: Physicochemical properties of some flotation reagents and their salts with ions of heavy non-ferrous metals. In "Proceedings of the Second International Congress on Surface Activity," Vol. 4. ed. J.H. Schulman. Butterworth, London, 1957, pp. 225-238.

CHAPTER 4

METHODS OF ANALYSIS OF XANTHATES

The quantitative determination of xanthates in solutions is often required in flotation plants. Aside from chemical methods based on certain well-known reactions of xanthate in solution, a few instrumental methods have been developed in recent years. The latter require the use of instruments such as the spectrophotometer, potentiometer, and polarograph. While these procedures are now widely employed in technical analysis, some of the chemical methods are still popular in laboratories when xanthates in high concentration are involved or when chemical analyses of the various constituents present in the admixture are required. The chemical methods presently in use will be described first.

CHEMICAL METHODS

IODIMETRIC METHOD

One of the oldest methods for the quantitative determination of xanthate is based on the quantitative oxidation of xanthate to diaxanthogen by iodine:

$$2S = C\begin{matrix} \diagup SK \\ \diagdown OR \end{matrix} + I_2 \longrightarrow S = \underset{\displaystyle OR}{\underset{|}{C}} - S - S - \underset{\displaystyle OR}{\underset{|}{C}} = S + 2KI \tag{4-1}$$

The analysis is done in aqueous medium. Cox[1] has recommended the use of 0.001 N iodine solution and the addition of a small quantity of potassium iodide to intensify the blue color at the end point using starch as indicator. For 20 ml of solution, 0.25 ml of 0.20 N potassium iodide and 0.2 ml of a 0.15% solution of starch are found to give satisfactory results. The indicator allowance is 0.1 ml. The use of stronger iodine solutions was not favored by Cox since he found it difficult to detect the blue starch iodide color, which according to him resulted from the slightly bluish color imparted to the solution by the dixanthogen formed as an emulsion. In our experience, however, there has been no serious difficulty in locating the end point even when 0.1 N solutions. In fact, accuracy within 1% has been claimed by working with 0.1 solutions.[2]

The iodimetric method of analysis has some serious limitations. It cannot be used for the determination of xanthate in the presence of reducing ions such as sulfide, sulfite, or thiosulfate. Certain metallic ions such as lead also interfere and give rise to erroneous results. The method is, therefore, useless except as a very rough guide when determination of xanthate is desired in flotation liquors and industrial effluents containing oxysulfur compounds of the type mentioned above. It is currently used in laboratories mainly as a check to ascertain the purity of freshly prepared xanthates.

IODOMETRIC METHOD

Calcott and co-workers[3] described a method based on the precipitation of xanthate as a copper salt and decomposition of a latter with nitric acid. The copper in solution is then determined by the iodometric method:

$$4S = C\begin{matrix} \diagup SK \\ \diagdown OR \end{matrix} + 2CuSO_4 \longrightarrow Cu_2 \left[\begin{matrix} S-C-OR \\ \| \\ S \end{matrix}\right]_2 + 2K_2SO_4$$

$$+ \; S = \begin{matrix} C \\ | \\ OR \end{matrix} -S-S- \begin{matrix} C \\ | \\ OR \end{matrix} = S$$

$$2Cu^{2+} + 4KI \longrightarrow Cu_2I_2 + I_2 + 4K^+ \quad (4\text{-}2)$$

A solution consisting of a mixture of cupric sulfate and sodium tartrate was recommended by Calcott. This solution is prepared by dissolving 25 g of cupric sulfate crystals in 600 ml of hot water, cooling to room temperature, and adding 200 g of sodium tartrate. After the tartrate is completely dissolved, a 10% solution of sodium carbonate is added until the color of the reagent turns to dark blue, indicating an excess of sodium carbonate; the solution is then diluted to 1 liter. The thiosulfate solution is standardized by using a sample of pure copper as the primary standard. About 1-2 g of xanthate sample are dissolved in 300 ml of cold water in a 500 ml conical flask, and about 125 ml of copper sulfatetartrate reagent slowly added with agitation. The flask is stoppered and the mixture is allowed to stand with occasional shaking in an ice water bath for at least 1 h but not more than 3 h. The yellow precipitate is then filtered off into a Gooch crucible and washed thoroughly with cold distilled water. The crucible containing the precipitate is placed in a beaker, and 30 ml of a mixture of 2 parts concentrated nitric acid to 1 part water by volume are poured into it. The mixture is allowed to stand for 5 min, and then the crucible is rinsed and removed. The copper nitrate solution containing the oily products of decomposition of the xanthate is filtered, and the filtrate and washings returned to the original 500-ml conical flask. Concentrated sulfuric acid (about 15 ml) is added and the solution is boiled until the nitric acid is completely expelled and dense fumes of sulfur trioxide fill the flask. The flask is allowed to cool, 100-125 ml of distilled water are added, and the solution again cooled to room temperature. Concentrated ammonia solution is added until a slight excess is present. The excess ammonia is boiled off and the solution cooled to room temperature and acidified with 7 ml of glacial acetic acid. About 5 g of potassium iodide are dissolved in the solution and the iodine liberated is titrated with 0.1 N sodium thiosulfate using 10 ml of 0.1% starch solution as an indicator.

The results are not affected by the presence of dixanthogen and impurities such as thiosulfate, sulfite, or carbonate. The presence of sulfides or thiocarbonate is, however, undesirable and causes high results. The method is, therefore, of very limited applicability and has not been used extensively.

ACID-BASE METHOD

An acid-base method for the analysis of xanthate was first described by Hirschkind.[4] A dilute solution of xanthate is acidified with an excess of standard hydrochloric acid. The liberated xanthic acid decomposes within 10 min into alcohol and carbon disulfide, both neutral substances, after which the excess of acid is back-titrated with standard alkali. The acid consumed corresponds to the xanthate present:

$$S{=}C\begin{matrix}\diagup SK\\ \diagdown OR\end{matrix} + HCl \longrightarrow KCl + S{=}C\begin{matrix}\diagup SH\\ \diagdown OR\end{matrix}$$

$$S{=}C\begin{matrix}\diagup SH\\ \diagdown OR\end{matrix} \longrightarrow CS_2 + ROH \qquad (4\text{-}3)$$

The carbon disulfide and alcohol formed do not have any effect on dilute acids and alkalies and are of no significance. Sulfates do not interfere when decinormal acids and alkalis are used. Carbonate and sulfates can be completely removed by precipitating with an excess of a 10% barium chloride solution and subsequent filtration. Hydroxide ions are determined by an initial titration with decinormal acid using phenolphthalein as an indicator. The interference from sulfide and thiocarbonate is avoided by working with dilute (less than decinormal) solutions so that all hydrogen sulfide liberated remains in solution.

As known aliquot of the solution (about 0.1 N) is mixed with 10-20 ml of a 10% solution of barium chloride and any carbonate precipitated is filtered off. To the filtrate an excess (approximately 50 ml) of decinormal hydrochloric acid is added and the mixture allowed to stand for 10 min. Then 3 drops of methyl red indicator are added and the solution titrated against decinormal sodium hydroxide solution until the color changes to yellow; the alkali consumption is noted. Phenolphthalein is then added as an indicator and the titration continued until the appearance of a pink color, which takes place when one-half of the hydrogen sulfide is neutralized. From the two titer values, the concentration of xanthate as well as that of sulfide can be calculated. If the xanthate solution initially contains hydroxide, allowance should be made for this in the final titer value. The hydroxide present initially can be determined by finding out the volume of decinormal acid required to neutralize the solution remaining after the carbonate is separated out. Phenolphthalein indicator should be used.

The volumetric method is accurate within ± 1% and is convenient for xanthate analysis in chemical laboratories. A further extension of this method in conjunction with the iodimetric method described above allows the estimation of thiosulfate present in the solution to be made. The solution left after titration with alkali is acidified again and then titrated with decinormal iodine solution. The iodine consumption directly indicates the thiosulfate present. Sulfite and carbonate may be determined in the precipitate obtained with barium chloride. Two equal samples of xanthate are precipated with an excess of barium chloride solution. The two precipitates are filtered and washed. One of them is ignited at a low red heat to determine the weight of barium carbonate plus barium sulfite. The second one is dissolved in dilute acid to which a certain amount of decinormal iodine solution has been added. After complete solution is obtained, the excess of iodine is titrated back with decinormal thiosulfate solution. The iodine consumed corresponds to the amount of sulfite present, and by subtracting the weight of the sulfate from the weight of the precipitate the carbonate content is ascertained.

A slightly modified method to determine the amount of impurities such as sulfide and thiocarbonate has been described recently.[5] The method of determination of sulfide ion involves the quantitative precipitation of sulfide as zinc sulfide by sodium zincate from an alkaline solution of xanthate. Other anions remain in a solution. About 2 g of xanthate sample are diluted with 20 ml water and 5 ml of sodium zincate solution, prepared by mixing 50 ml of 10% zinc chloride solution with 50 ml of 20% sodium hydroxide solution, are added. The precipitate is first washed with 0.5% sodium hydroxide, then with water, and centrifuged for 5 min after each washing. The precipitate is mixed with an excess of standard acidulated iodine solution and after 15 min the excess iodine is titrated with thiosulfate. Determination of the total amount of sulfide and thiocarbonate ions involves the quantitative precipitation of sulfide and thiocarbonate in the form of zinc sulfide and zinc thiocarbonate by the addition of zinc chloride (or sulfate) solution to the alcohol-water solution of xanthate. About 2 g of xanthate sample are diluted with a minimum quantity of water, and 100 ml of ethanol together with 2-4 ml of zinc chloride solution are added. The precipitate is allowed to stand for 10 min, then separated on alcohol-moistened filter paper and washed several times first with ethanol and then with water. The precipitate and the filter paper are treated with standard acidulated iodine solution and the excess iodine titrated with thiosulfate after 15 min. The amount of thiocarbonatc can be calculated from the differences between the total amount of sulfide and thiocarbonate and the amount of sulfide alone.

ARGENTOMETRIC METHOD

Makens[6] has described a volumetric method for the determination of xanthate by taking advantage of the low solubility of silver xanthate. Silver nitrate solution is added in slight excess to xanthate solution with constant stirring. About 30 ml of 10% ferric nitrate are added as an indicator, and excess standard potassium thiocyanate solution is run in from a buret to produce a distinct color. The excess thiocyanate is back-titrated with silver nitrate until the red color disappears. It is advisable to wrap the beakers with brown paper in order to minimize the decomposition of silver salts.

Under ideal conditions, the silver nitrate method compares favorably with the other two volumetric methods. However, it has a serious disadvantage. If the silver nitrate added is in large excess, the yellow precipitate of silver nitrate turns blck because of a complicated side reaction investigated by Makens.[6] The development of a black color before the addition of thiocyanate indicates that the experiment is vitiated. As it is difficult to adjust silver nitrate in slight excess, particularly when the approximate concentration of xanthate in the solution is not known, the method is not often adopted for quantitative analysis of xanthate.

4. METHODS OF ANALYSIS OF XANTHATES

MERCURIMETRIC METHOD[7]

This method is based on the reaction between xanthate and dimethylamine with the formation of dimethyl xanthamide and the release of sulfide ion:

$$ROCSS^- + 2NH(CH_3)_2 \longrightarrow ROCSN(CH_3)_2 + S^{2-} + NH_2(CH_3)_2^+ \qquad (4\text{-}4)$$

The resultant mixture is now titrated against o-hydroxy-mercuribenzoate (HMB); the following reaction occurs:

$$2\,^-OOCC_6H_4HgOH + S^{2-} \longrightarrow {}^-OOCC_6H_4HgSHgC_6H_4COO^- + 2OH^- \qquad (4\text{-}5)$$

The indicator solution is made by dissolving a small amount of thiofluorescein in 1 N ammonia solution and adding some EDTA to slow down the oxidation of thiofluorescein. The blue color of the indicator is discharged at the end point.

The sample of xanthate is dissolved in 5 ml of 40% aqueous dimethylamine, heated for 30 sec on a steam bath, and titrated with 0.05 N HMB in the presence of thiofluorescein as an indicator until the blue color disappears. At this stage the reaction is not complete and the final end point is reached after 1-3 min. The method is well applicable for xanthates and xanthic esters. Accuracy is within 1%

PERCHLORIC ACID METHOD[8]

When xanthate is dissolved in glacial acetic acid it decomposes, releasing an equivalent of acetate ion. This can be titrated against perchloric acid in glacial acetic acid using crystal violet indicator. The reactions are:

$$HClO_4 + CH_3COOH \longrightarrow CH_3COOH_2^+ + ClO_4^-$$

$$ROCSSK + CH_3COOH \longrightarrow ROCSSH + CH_3COO^- + K^+$$

$$CH_3COO^- + CH_3COOH_2^+ \longrightarrow CH_3COOH + CH_3COOH \qquad (4\text{-}6)$$

About 0.2 g of xanthate is dissolved in 15-20 ml glacial acetic acid; 2 drops of crystal violet indicator are added and the solution is titrated with 0.1 N perchloric acid to a distinct blue color.

GRAVIMETRIC METHOD

The low solubility of lead xanthates in water is made use of for the gravimetric determination of xanthate. In this method, described by Majumdar,[9] a measured quantity of xanthate solution is treated with 5-10 ml of a 10% lead nitrate solution and the contents shaken for 2-3 min in a separating funnel. About 25 ml of benzene are then added and shaking continued for 3 min. Lead xanthate is soluble in benzene while xanthates of inorganic acids are not. The benzene layer is separated and the aqueous layer (containing lead salts other than xanthate) drained off. Benzene is evaporated out on a tared glass dish and the residual lead xanthate weighed. From the weight of lead xanthate precipitate, the amount of xanthate present in the solution can be calculated. Majumdar has suggested an empirical correction factor which allows for the slight solubility of xanthate in water and gives more reliable results.

SPECTROPHOTOMETRIC METHODS

The adsorption of light in the ultraviolet region by xanthate has given rise to a very elegant method for the determination of xanthate in solution. Bushell and Malnarich[10] first obtained the spectrum of potassium isopropyl xanthate in the region 2000-3500 Å. As shown in Fig. 4-1, xanthate shows

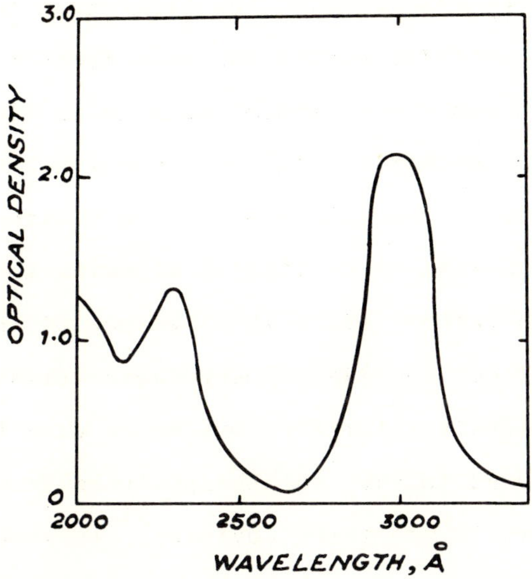

FIG. 4-1

Spectrum of a solution of potassium isopropyl xanthate [After C.H.G. Bushell and M. Malnarich: reproduced from Mining Eng., 1 (1956) by permission of the American Institute of Mining, Metallurgical and Petroleum Engineers.]

two absorption maxima, the one at 3010 Å being the more intense.

Intensity of absorption of light of this wavelength follows the Beer's law linear relationship between concentration and optical density and can be used for the spectrophotometric determination of xanthate. The concentration range that falls within the Beer's law plot has been shown to be 0-25 mg/liter. The method has been found to be extremely useful in xanthate analysis in flotation plants, for automatic reagent regulation systems, and in research laboratories. It is particularly useful for solutions containing reducing ions in high concentration, since the presence of these ions does not affect the light absorption. Dixanthogens do not seriously interfere, but since dixanthogen has been found to have an absorption maximum at 280-290 Å it is sometimes advisable[11] to separate them by extraction in a nonpolar solvent such as carbon tetrachloride.

ELECTROMETRIC METHODS

POTENTIOMETRIC METHODS

The action of silver nitrate on xanthate has been followed potentiometrically. The reaction may be written:

$$Ag^{+} + {}^{-}S\cdot CS\cdot OR \longrightarrow AgS\cdot CS\cdot OR \qquad (4\text{-}7)$$

During the precipitation the silver ion concentration (expressed as p Ag = $-\log\ (Ag^{+})$ is very low, changing in a regular manner, and increases suddenly when all the xanthate is precipitated. These changes in silver ion concentration measured with a silver electrode are followed in the potentiometric determination of xanthate first described by Du Rietz.[12,13] The experimental set up is shown in Fig. 4-2. The part of the salt bridge that is in contact with the sample contains saturated potassium nitrate or ammonium nitrate instead of

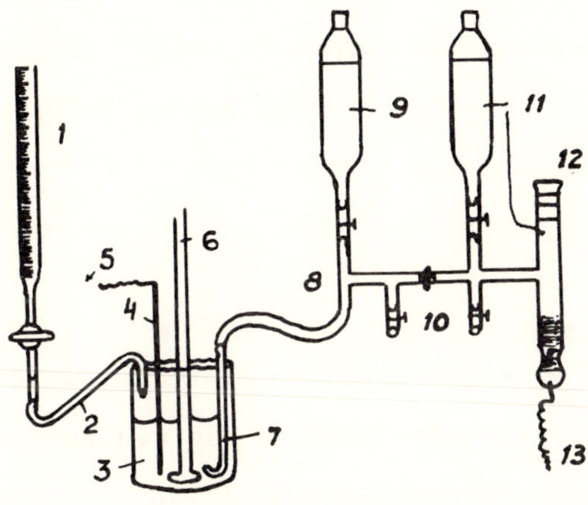

FIG. 4-2

Equipment for potentiometric titration of xanthates with silver nitrate. [After C. Du Rietz: Sartryck IVA, 24, 257-266 (1953).] (1) $AgNO_3$ Solution, (2,7) capillary glass tubes; (3) test solution; (4) silver thread; (5,13) to potentiometer; (6) glass stirrer; (8) salt bridge; (9) saturated KNO_3 solution; (10) glass cock; (11) saturated KCl solution; (12) calomel electrode.

potassium chloride. Before starting the titration a few drops of the nitrate solution are let out in order to ensure a clean contact surface. The silver electrode, which consists of pure silver wire about 1 mm thick, must often be wiped clean with a piece of emery cloth or cleaned with nitric acid. When the precipitate clusters during the titration and tends to cover the electrode surface, it may sometimes be necessary to shake the electrode. The addition of silver nitrate must be made drop by drop, especially during the early stages of the titration before the precipitation of the silver xanthate has gone to completion. The solution should be stirred each time before taking the emf reading.

The potentiometric method is useful for the analysis of xanthate samples containing thiosulfate. The reactions then would be:

$$Ag^+ + S_2O_3^{2-} \longrightarrow (AgS_2O_3)^-$$

$$Ag^+ + (AgS_2O_3)^- \longrightarrow Ag_2S_2O_3\ (s)$$

$$Ag_2S_2O_3 + H_2O \longrightarrow Ag_2S + H_2SO_4$$

(4-8)

The reactions, especially the last one, proceed rather slowly and during the later parts of the titration the potential drifts. In practice, however, it is possible to pass over the unstable region and reach a quite distinct end point before the final rise in emf caused by an excess of silver. The general nature of the potentiometric titration curves in shown in Fig. 4-3.

Strongly alkaline solutions of xanthate cause substantial errors in the determination of xanthate by the potentiometric method, probably as the result of secondary decomposition of the precipitated silver xanthate. These errors can be avoided by adding (before titration) sufficient boric acid to the alkaline sample solution to buffer it toward the strong alkali.

The use of cupric sulfate in place of silver nitrate in the potentiometric titration of xanthate was described by Oktawiec.[14] The solution of xanthate was placed in a beaker in which a calomel electrode and a copper electrode are inserted. The solution is titrated with 0.02 N cupric sulfate or 0.05 N cupric nitrate solution until reversal

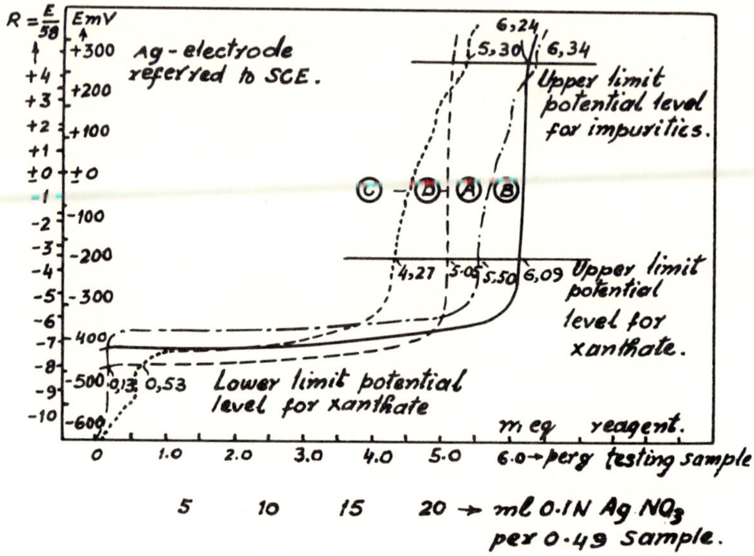

FIG. 4-3

Potentiometric titration curves for different xanthate samples with 0.1 M $AgNO_3$.
(A) Potassium ethyl xanthate, technical grade; (B) purified ethyl xanthate; (C) potassium amyl xanthate, technical grade; (D) purified amyl xanthate. [After C.Du Rietz: Sartryck Svensk Kem. Tidskri, 69, 310 (1957)]

of the electric current takes place. The copper electrode is made of electrolytic copper in the form of a plate. A blank titration is first carried out with a new copper electrode, since otherwise a sharp titration end point for the solution is not obtained. After a number of titrations the electrode must be cleaned with abrasive paper and a blank titration repeated.

This method is suitable for the determination of xanthates even in contaminated products. If any sulfide is present, it is removed by precipitation with a freshly prepared suspension of zinc carbonate. Carbonates and thiocarbonates are eliminated by adding a 20% solution of barium chloride. Accuracy of the method is within 0.5%.

CONDUCTOMETRIC METHOD

By measuring the conductivity of xanthate solutions of concentration 3-11% in the temperature range 10°-40°C, Oktawiec[15] established the following relations which can be used for the quantitative determination of xanthates.

$$XoNaBuX \times 10^{5} = 250 + 160C + (5 + 6.7C)t$$
$$XoNaEtX \times 10^{5} = 150 + 200C + (10 + 7C)t$$
$$XoNaAmX \times 10^{5} = 200 + 107C + (6 + 5.5C)t$$

where Xo denotes specific conductance, C the concentration, and t the temperature of the solution. Obviously these relations will not hold in the presence of other electrolytes, and empirical relations have to be determined for different systems. As such the method is of very limited applicability.

POLAROGRAPHIC METHOD

The polarographic potential of potassium ethyl xanthate (10^{-3} M) was found to be -0.29 V VS. SCE. This formed the basis of a polargraphic method of analysis by Cap and associates,[16] who recommended the use of 0.1 M nitrate, sulfate, or perchlorate as the supporting electrolyte and the presence of pyridine to avoid formation of insoluble films. The method was also described by Sun and Holzmann,[17] who employed it for the determination of xanthate in flotation liquor. It was found that the polargraphic wave of potassium ethyl xanthate at -0.26 V. vs. SCE was preceded not only by a prewave, probably attributable to the formation of a xanthate film on the mercury surface, but also by a sharp minimum at -0.06 V resulting from the anodic dissolution of mercury. As a result the calibration curve was rough and rather limited in concentration as shown in Fig. 4-4.

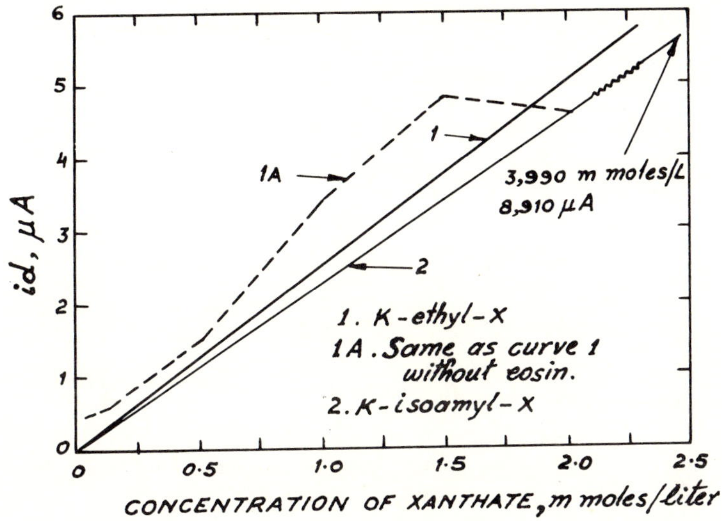

FIG. 4-4

Calibration curves for polarographic determination of purified xanthates.

[After S.C. Sun and R.T. Holzmann: Anal. Chem., 29, 1298, 1957; copyright 1957 by the American Chemical Society and reproduced by permission.]

The use of eosin eliminates this difficulty since it is preferentially adsorbed at the mercury-water interface, thus preventing the formation of an adsorbed film of xanthate. By adding 10^{-3} M eosin a linear relationship between the diffusion current (id) and the concentration of xanthate was obtained. The polarograms shown in Fig. 4-5 reveal that in the presence of eosin the prewave disappears.

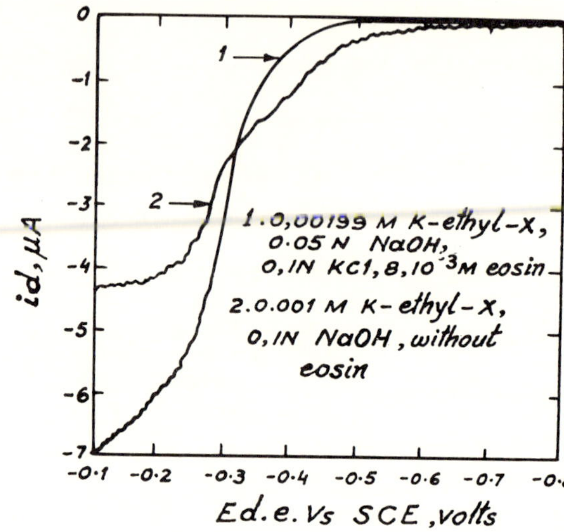

FIG. 4-5

Current voltage curves of purified potassium ethyl xanthate with and without eosin

[After S.C. Sun and R.T. Holzmann: Anal. Chem., 29, 1298 (1957); copyright 1957 by the American Chemical Society and reproduced by permission.]

The polarographic determination is carried out using a solution in which the xanthate concentration is within the range of 0.05-2.5 mmoles per liter in addition to 0.1 N KCl (supporting electrolyte), 0.05 N sodium hydroxide (pH regulator), and 10^{-3} M eosin. Oxygen-free nitrogen gas is passed through the solution for 13 min and a polarogram is obtained anodically over the range -0.1 to -0.7 V vs SCE. The concentration of xanthate is obtained by reading directly off the calibration curve.

This method has been checked for potassium isoamyl xanthate and the results show that it can be employed for analyzing all water-soluble alkali metal xanthates. Increasing the concentration of alkali does not affect the determination. The possible impurities such as sodium sulfide, carbonate, sulfite, silicate, and sulfate do not interfere. Thiosulfate does not affect the diffusion current for solutions containing up to 160 mg of potassium ethyl xanthate per liter as long as the ratio of xanthate to thiosulfate exceeds 4. This restriction is not a serious drawback for the analysis of xanthate in flotation liquors. Comparison with other established methods of analysis shows that the polarographic method is accurate within ±3%, which is considered satisfactory for the ordinary analysis of flotation liquors and research work on sulfide flotation.

RADIO TRACER METHODS OF ANALYSIS

Gaudin and Carr[18] have described a method for the radio tracer determination of xanthate employing ^{35}sulfur, which is a low-energy beta emitter with a half-life of 87.1 days. The activity can be measured by a gas flow counter or end-window-type

Geiger-Müller counter. The former has been found to be superior. The radioactive xanthate (carrying ^{35}S) is oxidized in an alkaline solution of permanganate and the resultant sulfate is precipitated as barium sulfate. The observed activity from a precipitate made with a standard amount of ^{35}S is inversely proportional to the weight of the precipitate. Gaudin and Carr recommend the following procedure. To an aliquot of radioxanthate solution containing 0.1-1 mg of xanthate, excess potassium permanganate (3 ml of solution containing 16 g of potassium permanganate, and 10 g of potassium hydroxide per liter)is added, and the solution heated gently and carefully to dryness. Concentrated hydrochloric acid is added to decompose the potassium permanganate and manganese dioxide, and the solution evaporated to dryness. The residue is dissolved in 400 ml of water, and 1 ml of concentrated hydrochloric acid and carrier sulfate solution to produce about 100 mg of barium sulfate precipitate are added. The solution is heated to boiling. About 25 ml of a 5% solution of barium chloride are slowly added with constant stirring. The precipitate is allowed to digest for at least 1 h and to stand overnight. It is then transferre to a filtration apparatus and washed with alcohol and acetone. The precipitate and filter paper are mounted on a brass disk for radiometric determination.

The accuracy of the tracer method is ±0.6%. The technique is particularly useful and has been extensively used for the determination of xanthate content in the adsorbed layers on mineral surfaces. The adsorbed layer is leached by about 250 ml of hot pyridine per 5-10 g of sulfide mineral. In each case concentrated hydrochloric acid is added to neutralize the pyridine after oxidizing the xanthate sulfur by the addition of potassium permanganate. The method is of potential utility in studying various aspects of mineral-xanthate systems. These include: (1) determination of the partition of xanthate between solution and mineral, and ascertainment of the reversibility or nonreversibility of the xanthate-mineral system; (2) investigation of the factors that determine the rate at which xanthate is adsorbed by the mineral, and (3) determination of the effect of other ions in solution on xanthate adsorption. Some of these applications are described in Chapter 7.

4. METHODS OF ANALYSIS OF XANTHATES

ANALYSIS OF TECHNICAL XANTHATES

The analytical techniques outlined in the preceding sections can in many cases be adopted for the analysis of technical xanthates. For samples that contain impurities in high percentage, however, some special methods have been developed for analysis. Kilmenko and co-workers[19] have described a chemical method in which the impurities are separated and analyzed. The procedure recommended is as follows: 1 g of the sample in 100 ml water is treated with 10-30 ml of freshly prepared zinc carbonate suspension. The mixture is allowed to stand for 10 min (solid xanthate samples) to 2 h (solutions) and then filtered into a volumetric flask. The precipitate contains sulfide and trithiocarbonate. To determine the concentration of these compounds the precipitate is titrated with 10 ml of water and 10 ml of 0.5 N iodine solution and then treated with 5 ml of 2 N hydrochloric acid. After the solution stands for 10-15 min in darkness, the excess iodine is titrated with thiosulfate to determine the sum of sulfide and thiosulfate in the xanthate sample. The filtrate contains not only the xanthate but also the thiosulfate and the sulfite in the sample. To a 50-ml aliquot, 1 ml of 0.1% sodium rhodizonate and 1-2 drops of formaldehyde are added, and the solution is titrated with a 0.025 N lead acetate solution to a pink-violet-color end point. The solution of lead acetate should be standardized against a pure sample of xanthate. Another 100-ml aliquot is treated with 2.5 ml of 20% barium chloride solution, stirred, and after being allowed to stand for 5-10 min is filtered. The filtrate is treated with 20 ml of standard (0.1-0.5 N) hydrochloric acid and allowed to stand for 10-30 min, to decompose the xanthate. It is then titrated with iodine to determine the thiosulfate content. To determine sulfate, sulfite, and carbonate, two samples of 1 g each in 50 ml of water are treated with 10 ml of 10% barium chloride solution. The precipitates are filtered and washed. The first precipitate is tritrated with 10 ml of 0.05 N iodine and 5 ml of 2 N hydrochloric acid and after 10-15 min the excess iodine is titrated with sodium thiosulfate solution. The second precipitate is washed with 1 N hydrochloric acid 3-4 times and then with water 5-6 times. The residue is barium sulfate. The carbonate content is calculated by difference, making allowance for sulfite, which also passes into solution on treatment of the original precipitate with hydrochloric acid along with barium carbonate.

Aside from the impurities mentioned above, xanthate samples also contain moisture and it is sometimes necessary to make an estimation of moisture content. This is a little complicated since rapid heating of the sample causes an extensive decomposition accompanied by an abnormal loss of weight. If the heating is slow, however, the water can be removed from the xanthate without chemical changes. In the dry state xanthates remain stable at a temperature of 100°C for a fairly long time. The simplest procedure to determine the moisture content, therefore, is to weigh out 1.00 g of the sample in a weighing glass which is then put into a drying oven at room temperature. The oven temperature is slowly increased to 100°C. When this temperature is reached, the sample is withdrawn, cooled, and weighed. The loss of weight gives the moisture percentage of the sample.

An accurate method of determining moisture in technical xanthates using Karl Fischer reagent has been described by Linch.[20] The reagent is prepared by dissolving 404 ml of pyridine in 1000 ml of anhydrous methanol and adding 127 g of iodine crystals. When dissolved, the solution is chilled in ice, and 100 g of sulfur dioxide are added as a gas under the surface of the liquid. Standardization is done against standard water solution in methanol. In the procedure recommended, 25 ml of chloroform are measured into a dry 125 ml Erlenmeyer flask and Karl Fischer reagent is added dropwise through a two-hole stopper until a definite visual point is reached (2-3 ml) with vigorous swirling to wash down sides of the flask. The xanthate sample (about 0.5-5 g) is quickly added, and the solution titrated to a permanent end point with constant swirling. Samples containing more than 10% moisture should be reduced to 0.5-1.0 g in size, while less than 1% moisture requires a 4.0-5.0 g samples for accuracy. The percentage of water is given by the expression

$$\frac{\text{Millliliters of K.F. reagent X factor (as grams of } H_2O \text{ per milliter) X 100}}{\text{Sample weight}}$$

The average accuracy of this method is $\pm$ 1%. The presence of thiosulfate vitiates this analysis, but sulfide and sulfite do not affect the accuracy.

In technical analysis it is often necessary to determine not only the xanthate percentage but also the molecular weight of the sample in order to identify the hydrocarbon chain. The molecular weight of the sample can be determined after purification of a small quantity and analysis by any suitable method. If the molecular weight is M and the atomic weight of the metal atom is denoted by Me, the number of carbon atoms n is given by

$$n = 1/14(M - 93 - Me)$$

(where 14 is the atomic weight of carbon and 93 is the molecular weight of xanthate group minus the carbon atoms.) If there is any uncertainty as to whether the metal atom is sodium or potassium, it is possible to determine the atomic weight of the metal by burning 1 g of the purified portion of the sample in a platinum crucible and weighing the ash. In order to obtain the latter in a defined form (as pure sulfate), a few drops of sulfuric acid are added and the combustion accomplished at 600-700 °C. Du Rietz[13] has described the details of analysis of technical xanthate by the above principle and by making use of potentiometric titration technique described earlier.

The resulting ash is weighed. In this process xanthate is completely decomposed and the residual metal is converted into the corresponding sulfate. As 1 equivalent of xanthate gives 1 equivalent of sulfate, the number of equivalents of sulfate must be equal to the number of equivalents of xanthate. The number of equivalents of xanthate is estimated by titrating a known weight of the sample against iodine. If g is the weight of the sulfate ash obtained, a the titer value in milliters of 0.1 N iodine solution, and w the weight in grams of the xanthate taken for titration,

$$\frac{g}{48 + Me} = \frac{a \times 0.1}{1000 \times w}$$

Equivalent weight (Me) of the metal atom of xanthate can be calculated. In combination with the potentiometric technique described above, this method is recommended for the analysis of technical xanthate samples.[13]

REFERENCES

1. Cox, A.B.: Titrations with 0.001 N iodine. Soc. Chem. Ind. (Victoria), 35 996 (1935); 35, 1001 (1935).

2. Matuszak, M.P.: Iodometric determination of carbon disulphide. Ind. Eng. Chem. (Anal. Ed.), 4, 98-102 (1932).

3. Calcott, W.S., F.L. English, and F.B. Downing: Analysis to xanthate. Eng. and Mining J. Press 118, 980 (1924).

4. Hirschkind, W.: Rapid method of analyzing xanthates. Eng. Mining J. Press 119, 968 (1925).

5. Oktawiec, M., I. Wojtowicz, and K. Zmudzinski: Determination of certain impurities in technical xanthates. Prace Inst. Hutnicych, 12 (1), 1-7 (1960).

6. Makens, R.F.: The silver xanthates. J. Am. Chem. Soc., 57, 405 (1935).

7. Veibel, S., and M. Wronski: Mercurimetric estimation of alkyl xanthates. Acta Chem. Scand., 20, 849-852 (1966).

8. Berger. J.: Determination of xanthogenates by titration with perchloric acid. Acta Chem. Scand., 6, 1564-1567 (1952).

9. Majumdar, K.K.: New method for the estimation of xanthates. J. Sci. Ind. Res. (India), 11B, 260 (1952).

10. Bushell, C.H.G. and M. Malnarich: Reagent control in flotation. Mining Eng., 8, 734-737 (1956).

11. Suvorovskaya, N.A. and V.V. Shikhova: Extraction spectrophotometric method for the determination of dixanthogen and potassium xanthate. Kontrol Raspredeleniya Ksontogeneta v Protsesse Flotatsii, Akad. Nauk SSSR, Inst. Gorn.Dela (Chem.Abstr., 65, 11, 347, 1966).

12. Du Rietz, C.: Chemical problems in the flotation of sulphide minerals. Sartryck ur IVA, 24, 257-266 (1953)

13. Du Rietz, C.: Xanthate analysis by means of potentiometric titration. Sartryck ur Svensk Kem. Tidskr., 69, 310-327 (1957).

14. Oktawiec, M.: Potentiometric method for the determination of xanthates. Prace Inst. Hutniczych, 6, 184 (1954); Chem. Abstr., 49, 5212d (1955).

15. Oktawiece, M.: Determination of the concentration of several xanthates by their specific conductance. Bergakademie, 18(4), 239-244 (1966); Chem. Abstr., 65, 1836 (1966).

16. Cap, M., A. Dvork, and V. Hencl: Polarographic and potentiometric determination of xanthates. Sbornik Celostatni Pracovni Konf. Anal. Chemiku, 1st, Prague, 1952, 174-179 (1953); Chem. Abstr., 50, 3160 (1956).

17. Sun, S.C. and R.T. Holzmann: Polarographic determination of xanthate in flotation liquors. Anal. Chem., 29, 1298-1301 (1957).

18. Gaudin, A.M. and J.S. Carr: Analytical procedures for ethyl xanthate labeled with radiosulphur. Anal. Chem., 24, 887-888 (1952).

19. Klimenko, Yu V., I.A. Kakovskii, E.I. Silina, and C.P. Anufrieva: Analysis of technical xanthates. Zavodk. Lab., 14, 1410 (1948).

20. Linch, A.L: Determination of purity and water contents of xanthates and dithiocarbamates. Anal. Chem., 23, 293-296 (1951).

CHAPTER 5

DIXANTHOGENS

As mentioned in Chapter 3, dixanthogens are the oxidation products of xanthates. They have the general formula

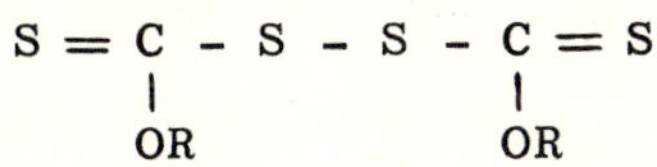

They are also called by other names such as o-alkylthiocarbonic acid disulfide, bisalkyl xanthogen, and 0-0-dialkyl ester of bisthioformic acid. (The latter term is used in Chemical Abstracts).

The quantitative reaction between potassium xanthate and iodine in aqueous solution can be employed to prepare dixanthogen. A white turbid liquid will be obtained, the thick turbidity is a result of the dispersion of the dixanthogen in water. It is extracted in ether or petroleum ether by vigorous shaking. The ethereal layer is separated and the dixanthogen is isolated by evaporating off the solvent by applying a mild vacuum.

Other oxidizing agents such as chlorine or sodium hypochlorite may be employed. The oxidation of xanthate by cupric ions in solution also yields dixanthogen, which can be separated from cuprous xanthate by extraction with ether. Electrolytic oxidation of xanthate using a platinum anode has also been described.[1]

Preparation of a series of dixanthogens by the oxidation of xanthate ion by nitrous acid has been reported by Gurvich and Belova.[2] Carbon disulfide, alcohol, and sodium hydroxide (concentrated aqueous solution) are mixed in equimolar proportions, and the temperature is maintained below 30°C. After 30 min sodium nitrate is added in slight excess. The temperature is lowered to -3°C and 25% sulfuric acid is added. The dixanthogen separated is purified by extraction with ether.

Dimethyl dixanthogen is obtained as an oily liquid. It decomposes on distillation, forming the methyl ester of methyl xanthic acid. Diethyl dixanthogen is also oily when it is prepared by extraction with ether, but it can be crystallized by cooling. The solid melts at 28°C. Thermal decomposition yields the ethyl ester of ethyl xanthic acid and other products in small quantities such as carbon disulfide and sulfur. In ether it combines directly with potassium or sodium in the presence of a trace of alcohol to form ethyl xanthate. With alcoholic potash it forms potassium ethyl xanthate, carbon dioxide, and sulfur. The action of alcoholic potassium hydrosulfide on diethyl dixanthogen yields hydrogen sulfide and sulfur, and potassium ethyl xanthate remains in solution. If the alcoholic solution is saturated with ammonia, sulfur separates out and the solution contains the ethyl ester of monothiocarbamic acid ($O=C(NH_2)(SC_2H_5)$) and ammonium ethyl xanthate. Exact mechanisms of the reactions have not been investigated although the qualitative observations were described more than 100 years ago.[3,4]

Dipropyl dixanthogen is a liquid (density, 1.9661 g cm^{-3}).[5] It decomposes at about 150°C presumably by steps similar to those mentioned above. Diisobutyl dixanthogen occurs as a heavy yellow oil which solidifies below -10°C. Diisoamyl dixanthogen is also obtained as a yellow oil. The chemical reactions are similar to those of diethyl dixanthogen.

Gurvich and Belova[2] have determined physical constants such as melting point, boiling point, and density of a series of dixanthogens. Their data are reproduced in Table 5-1.

TABLE 5-1

Physical Constants of Dixanthogens[a]

Dixanthogen	Melting point, °C	Boiling point, °C	Density at 20°C g cm^{-3}	Refractive index (n_D^{20})
Dimethyl	22.5-23	104-106 (at 0.5 atm)	-	-
Diethyl	31.5	107-188 (at 0.2 atm)	-	-
Dibutyl	-	134-135 (0.35 atm)	1.1518	1.5800
Dihexyl	-	149-152 (0.35 atm)	1.0897	1.5711

[a] From S.M. Gurvich, and R.Ya. Belova: Zh. Obshch. Khim., 31, 1631-1635 (1961).

GENERAL CHARACTERISTICS OF DIXANTHOGENS

Only the lower dixanthogens up to dihexyl dixanthogen have been studied in some detail. As revealed from Table 5-1, they are all oily liquids with a mild odor. As is to be expected from their nonpolar structure, they are comparatively inert and very few chemical reactions are known. These reactions have been used in their quantitative determination and are described in the next section.

Dixanthogens are readily reduced to xanthate by nascent hydrogen generated in alkaline medium by the action of a metal such as zinc

$$S{=}\underset{\substack{|\\OR}}{C}{-}S{-}S{-}\underset{\substack{|\\OR}}{C}{=}S + 2H + 2NaOH \longrightarrow 2S{=}C\begin{matrix}\diagup SNa\\ \diagdown OR\end{matrix} + 2H_2O \qquad (5\text{-}1)$$

ANALYSIS OF DIXANTHOGENS

Two chemical methods have been described for the quantitative determination of dixanthogens. The first is based on the oxidation of dixanthogen by chloramine-T in an acidic medium. It has been shown[6] that 1 mole of diethyl dixanthogen consumes 26 equiv of chloramine-T according to the following mechanism:

$$\underset{\displaystyle OC_2H_5}{S = \overset{|}{C}} - S - S - \underset{\displaystyle OC_2H_5}{\overset{|}{C}} = S \quad + \quad 13O + 7H_2O \longrightarrow$$

$$2C_2H_5OH + 4H_2SO_4 + 2HCOOH \tag{5-2}$$

which is similar to the mechanism proposed for the oxidation of ethyl xanthate. For actual analysis a known volume of dixanthogen solution is mixed with an excess of acidified chloramine-T (10 ml of 0.1 N with 5 ml of 2 N hydrochloric acid), and the mixture set aside for 1 h. Then, 10 ml of 10% potassium iodide are added and the liberated iodine is titrated against a standard solution of thiosulfate. Reactions between higher dixanthogens and chloramine-T are more complicated, consuming only 18-19 equivalents of chloramine-T instead of 26 equivalents. The quantitative determinations should, therefore, follow calibration using solutions containing known amounts of dixanthogens in the same concentration range as that of the unknown solution.

The second method, described by Patel and Shankarnarayana,[7] is based on the reaction between dixanthogen and potassium cyanide first described by Whitby and Greenberg.[8] Xanthogen monosulfide is formed as a product of reaction.

$$S{=}C(OR){-}S{-}S{-}C(OR){=}S + KCN \longrightarrow S{=}C(OR){-}S{-}C(OR){=}S + KCNS \tag{5-3}$$

Thiocyanate is allowed to react with bromine to form cyanogen bromide in accordance with the equation

$$SCN^- + 4Br_2 + 4H_2O \longrightarrow CNBr + 7Br^- + SO_4^{2-} + 8H^+ \tag{5-4}$$

The cyanogen bromide is next treated with potassium iodide, which is oxidized, forming iodine.

$$CNBr + 2I^- \longrightarrow CN^- + Br^- + I_2 \tag{5-5}$$

The iodine is then titrated with standard thiosulfate solution.

5. DIXANTHOGENS

Ten milliliters of dixanthogen solution are mixed with 5 ml of 10% ammonium nitrate solution (which helps in the extraction of xanthogen monosulfide by benzene), 10 ml of 2% potassium cyanide solution, and 10 ml of acetone. The mixture is heated at 40°-50°C for about 25 min to ensure completion of the reaction between the dixanthogen and cyanide. The mixture is cooled and shaken well with 15 ml of benzene in a separating funnel to extract the xanthogen monosulfide formed. The extraction should be repeated until the monosulfide is completely removed (as indicated by the disappearance of the yellow color of the solution). The aqueous solution is then mixed with 10 ml of 17% orthophosphoric acid and heated on a steam bath to remove hydrocyanic acid. The solution is cooled, and bromine water is added dropwise until bromine is in slight excess. The excess of bromine is removed by adding small amounts of ferrous sulfate until the yellow color disappears. Potassium iodide and a little sodium bicarbonate are added, and the mixture is vigorously shaken and set aside in the dark for 10 min. The liberated iodine is titrated with standard sodium thiosulfate solution using a starch indicator.

In conjunction with an acid-base titration method for the determination of xanthate (see Chapter 4), this method has been successfully employed for the determination of xanthate and dixanthogen separately when they are present together in an aqueous solution. A solution of xanthate kept in atmosphere for an appreciable length of time can thus be analyzed. The solution is first analyzed for xanthate. About 15-20 ml of acetone are added to the solution remaining after acid-base titration in order to clarify it, and the dixanthogen is determined by the cyanide method.

Determination of dixanthogen can also be accomplished by spectrophotometric methods although there is some disagreement in literature regarding the exact absorption maxima of dixanthogens. Suvorovskaya and Shikhova[9] found that the absorption maximum of dibutyl dixanthogen occurs at 280-290 mμ, which is rather close to the value for xanthate (mentioned in Chapter 4). Because of the closeness of peaks, if the two compounds are present together the dixanthogen has to be separated by extraction with carbon tetrachloride and then estimated by its abosrbance at 290 mμ.

Pomianowski and Leja[10] found two wide absorption bands for dixanthogen, one occurring between 280 and 300 mμ and the other between 220 and 250 mμ. In a nonpolar solvent such as ether, the bands are more clearly defined at 286 and 241 mμ (Fig. 5-1). Apparently the polar character of the solvent influences the absorption. It is not known whether this is the result of an interaction with the solvent molecules.

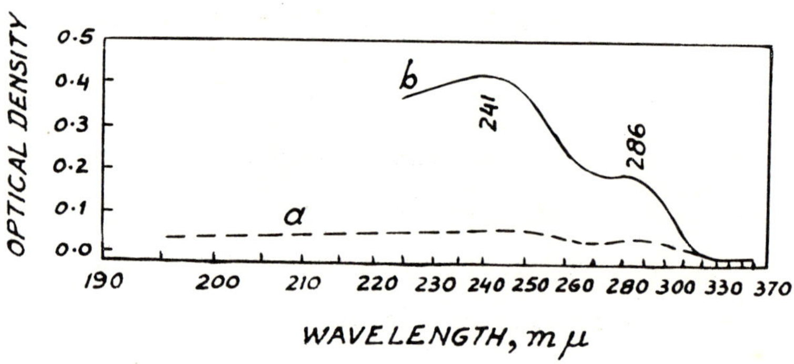

FIG. 5-1

Ultraviolet spectra of diethyl dixanthogen. (a) Freshly prepared aqueous solution in distilled water; (b) 4.1 X 10^{-3} M solution in diethyl ether. [After A. Pomianowski and J. Leja: Can. J. Chem., 41, 2219 (1963); reproduced by permission of National the National Research Council of Canada.]

Polarographic determination of dixanthogen can be carried out as described by Plaksin and Shrader[11] in aqueous alcoholic solution with 0.1 N sodium acetate as the supporting electrolyte. The average half-wave potential for a 4 X 10^{-5} M solution is found to be -0.42 V vs. SCE. The method is rapid and the maximum relative error is 5%.

THE XANTHATE-DIXANTHOGEN SYSTEM

As described in Chapter 3, the oxidation of xanthate to dixanthogen in aqueous solution has been the subject of several investigations. Because of its major role in the flotation of sulfide minerals (see Chapter 7), the xanthate-dixanthogen system has been analyzed by spectrophotometric and potentiometric techniques.

The decomposition of xanthate in an aqueous solution containing dixanthogen was studied by Pomianowski and Leja[10] who identified the individual species formed by spectrophotometric methods based on the absorption maxima at different wavelengths for products of decomposition such as dixanthogen and carbon disulfide. Ethyl xanthate ion has a characteristic absorption maximum at 3010 Å; diethyl dixanthogen shows a peak at 2410 Å and carbon disulfide at 2065 Å. The authors have found that the optical densities in the region below approximately 2800 Å are affected by the length of the storage period since recrystallization of the solid xanthate and also by the method of preparation of the solutions. Spectra of dilute solutions (10^{-5}-10^{-4}M) obtained by dissolving the required quantity of xanthate stored for a prolonged time shows higher optical densities in the region below 2800 Å than those of the solutions obtained by dissolution of a freshly recrystallized sample. This variation in optical density has been examined further. A solution of aged xanthate adjusted to pH 8.5 and kept for 19 h at 85°F shows an increase in optical density at 3010 Å as a result of the formation of xanthate ion from dixanthogen originally present at the surface and an increase in optical densities below 2160 Å caused by the formation of carbon disulfide. The reduction of dixanthogen to xanthate has also been confirmed with a freshly made solution of ethyl xanthate kept at 98°F. As shown by the spectrum (Fig.5-2), the absorption maximum of dixanthogen falls with time

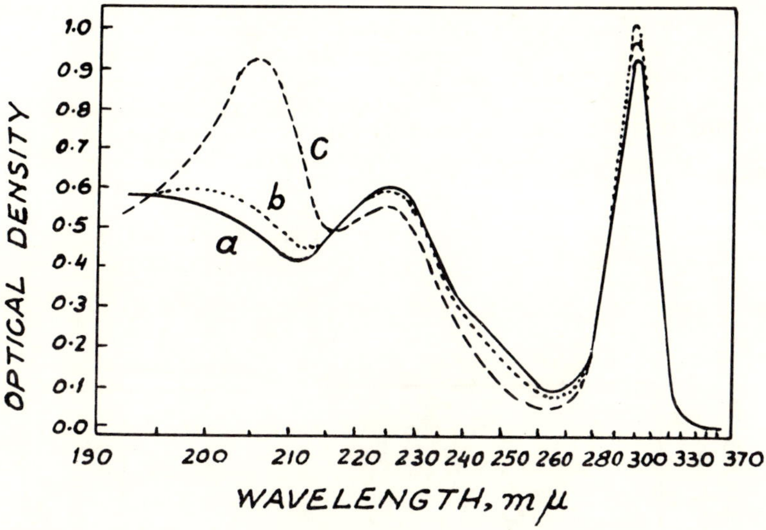

FIG. 5-2

Ultraviolet spectra of 1.1 X 10^{-4} M ethyl xanthate solution at pH 6 kept in a closed cell at 98°F for (a) 0 h, (b) 7 h, and (c) 24 h. [After A. Pomianowski and J. Leja: Can. J. Chem., 41, 2219 (1963); reproduced by permission of the National Research Council of Canada.]

and that of carbon disulfide rises. There is also a significant change in optical density at 3010 Å after 7 h, followed by a decrease after 24 h. This shows that under experimental conditions the decomposition of dixanthogen to xanthate is faster than its decomposition to carbon disulfide and alcohol. When distilled water is kept in contact with dixanthogen for several days, the resultant spectrum (Fig. 5-3) indicates the presence of all three species, xanthate ion, dixanthogen, and carbon disulfide. A steady state is reached after 15-20 days. The result of reduction is enhanced in a strongly alkaline medium (See Fig. 5-4).

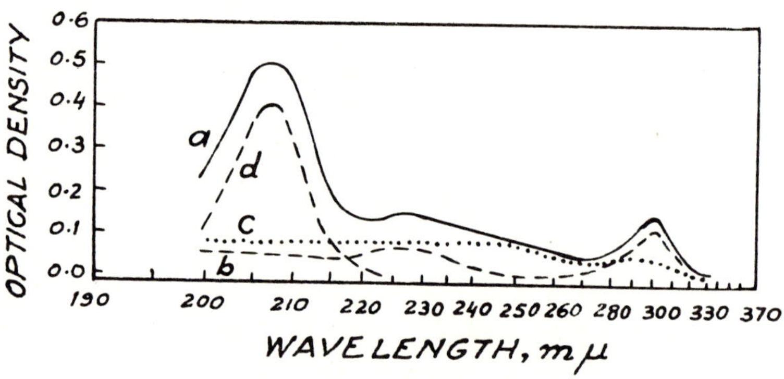

FIG. 5-3

Sequence of ultraviolet spectra indicating the products of dixanthogen decomposition. (a) Saturated dixanthogen, solution; (b) xanthate solution equivalent to the difference of (a) and (c); (c) freshly prepared dixanthogen solution; (d) CS_2 solution, difference of (a), (b), and (c). [After A. Pomianowski and J. Leja: Can. J. Chem., 41, 2219 (1963); reproduced by permission of the National Research Council of Canada.]

The results of Pomianowski and Leja[10] reveal the instability of dixanthogen and the possibility of six reactions governing the xanthate-dixanthogen system.

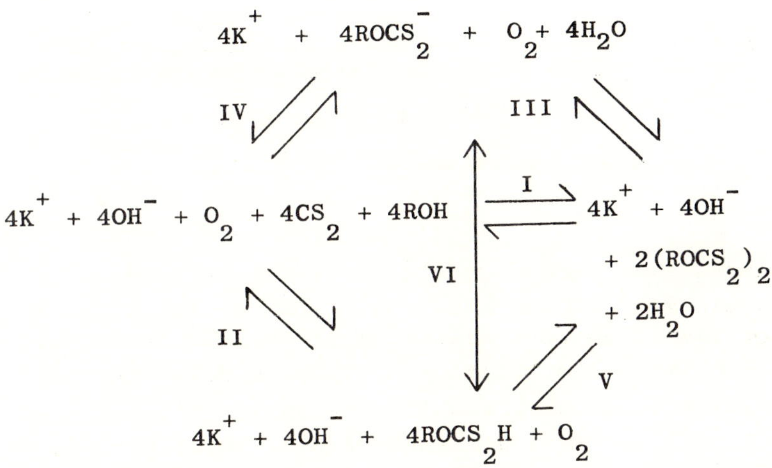

(5-6)

Reactions I, II, and III are characteristic reactions of xanthate and were discussed in Chapter 3. Reaction IV is employed in the preparation of xanthates (Chapter 2). The irreversibility of reaction III, which implies reduction of dixanthogen, is shown by spectrophotometric studies. The authors also evaluated, from spectrophotometric measurements of dixanthogen solutions, the solubility of diethyl dixanthogen in water (found to be

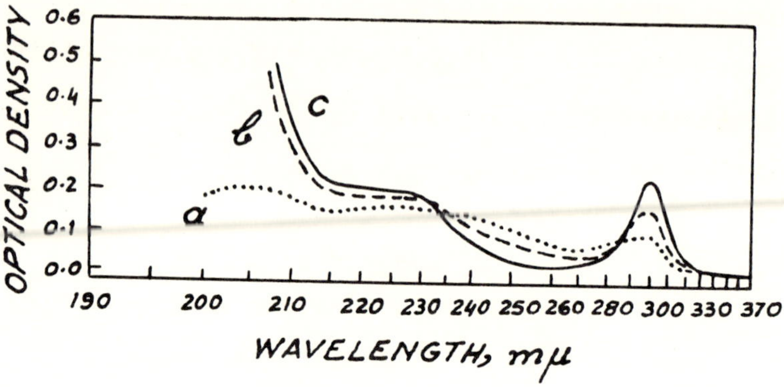

FIG. 5-4

Changes with time in the ultraviolet spectrum of dixanthogen aqueous solution at pH 10. (a) Initial spectrum of aqueous dixanthogen; (b) spectrum taken 15 min after the addition of NaOH to raise the pH to 10; (c) the same solution as in (b) after 2 h at 40°C. [After A. Pomianowski and J. Leja: Can. J. Chem., 41, 2219 (1963); reproduced by the permission of the National Research Council of Canada.]

1.3×10^{-5} mole/liter) and the equilibrium constant of this reaction, viz. $[EtX^-]^2/[OH^-]^2\,[(EtX)_2] = 3.3 \times 10^{10}$. This can be used to calculate the approximate concentrations of diethyl dixanthogen in xanthate solutions of different concentrations and at different pH values. It has been found that solutions of ethyl xanthate become saturated with respect to dixanthogen at different pH values, depending on the concentration of xanthate ion; for 10^{-3}, 10^{-4}, 10^{-5}, and 10^{-6} M concentrations of xanthate ion in solutions, the pH values of saturation are, respectively, 8,7,6, and 5.

The value of the equilibrium constant for the xanthate ion-dixanthogen system mentioned above is in complete disagreement with the values recognized by Finkelstein,[12] who determined the value of the standard oxidation potential of the dixanthogen-xanthate couple by an independent redox indicator method. The values he obtained lie between -0.08 and -0.14 V. They are in fair agreement with the values determined by other workers[13] for a homologous series of xanthates. Taking -0.07 V as the representative value for the standard oxidation potential, Finkelstein calculated the equilibrium constant of the reaction $4EtX + O_2 + 2H_2O \longrightarrow 2(EtX)_2 + 4KOH$ and obtained 6.6×10^{31}. This value implies that the equilibrium reaches far to the right and the reverse reaction, that is, reduction of dixanthogen to xanthate, is virtually ruled out. The cause for this serious discrepancy between the results obtained from two different techniques is not understood at present. Finkelstein has expressed the view that there may be an error in at least one of the following assumptions implicit in the treatment of spectrophotometric data: (1) the solution obtained from the contact of pure dixanthogen with water contains dissolved dixanthogen long enough for spectra to be determined: (2) the absorption maximum at about 2800 Å is characteristic of dixanthogen in aqueous solution: (3) the absorption maximum at 3010 Å is characteristic of xanthate in aqueous solution. There is sufficient evidence to support the third of these assumptions. Assumption (2), as mentioned above, is not sufficiently supported, while assumption (1) seems reasonable under the experimental conditions but lacks confirmation. Clearly the system deserves deeper chemical analysis to check the various possible modes of decomposition of dixanthogen that have been postulated.

Standard oxidation-reduction potentials for a homologous series of xanthate-dixanthogen couples have been determined by Kakovskii and associates.[13] It is found that an increase in the length of the hydrocarbon chain of xanthate increases the potential of the oxidation reaction. This implies that higher xanthates are stronger reducing agents than the lower derivatives. Partial reversibility of the reaction on a solid platinum electrode occurs for lower dixanthogens, but no direct proof of the reversibility has been put forward.

REFERENCES

1. Gomez, I: Isoamyl xanthates. Anales Soc. Espan. Fis Quim., 14, 91-102 (1916).

2. Gurvich, S.M. and R. Ya. Belova: Some derivatives of xanthic acids. Zh. Obshch. Khim. 31, 1631-1635 (1961).

3. Derechsel: Z. Chemi., 583 (1865).

4. Debus: Ann. Chem., A75, 122.

5. Scala, N.: Gazz. Chim. Ital., 17, 80. (References 3,4,5 have been cited from Beilstein's Handbuch der Organische Chemie, Vol. 3, p.214).

6. Rao, S.R.: A method for the determination of dixanthogen. Talanta, 8, 746-747 (1961).

7. Shankarnarayana, M.L. and C.C. Patel: The volumetric determination of dixanthogen. Analyst, 86, 98-101 (1961).

8. Whitby, G.S. and H. Greenberg: Dialkyl sulfurdicarbothionates. Trans. Roy. Soc. Can., 23, 21 (1929).

9. Suvorovskaya, N.A. and V.V. Shikhova: Extraction-spectrophotometric method for the determination of dixanthogen and potassium xanthate. Kontrol Raspredeleniya Ksantogenata V Protssesse, Flotatsii, Akad. Nauk SSSR, Inst.Gorn.Dela, 49-62 (1965); Chem. Abstr., 65, 10, 193 (1966).

5. DIXANTHOGENS

10. Pomianowski, A. and J. Leja: Spectrophotometric study of xanthate and dixanthogen solutions. Can. J. Chem., 41, 2219-2230 (1963).

11. Plaksin, I.N. and E.A. Shrader: Polarographic determination of dixanthates. Izv. Vysshikh Uchebn. Zavedenii Tsvetn. Met., 5(1), 41-43 (1962); Chem. Abstr., 57, 4039 (1962).

12. Finkelstein, N.P.: Kinetic and thermodynamic aspects of the interaction between potassium ethyl xanthate and oxygen in aqueous solution. Trans. Inst. Mining Met., 76, C51-C59 (1967).

13. Kakovskii, I.A, N.V. Sebrayakova, and B.A. Stepanov: Oxidation reduction potentials of xanthates. Nauch. Doklady Vysshei Shkoly, Khim. Khim. Teknol., No.2, 277-279 (1959); Chem. Abstr., 54, 12840 (1960).

CHAPTER 6

STRUCTURAL CHEMISTRY OF XANTHATES AND DIXANTHOGENS

The structural chemistry of xanthates and dixanthogens was extensively investigated during the 1960's by employing infrared techniques, X-ray crystallography, and the measurement of dipole moments and magnetic susceptibilities. The data so obtained have provided valuable information on the nature of the bonds that comprise the molecule and the distribution of electron density. Sometimes the data are also useful in elucidating the reactivity of different xanthates.

INFRARED SPECTRA OF XANTHATES AND DIXANTHOGENS

Felumb[1] first obtained the infrared spectra of dixanthogens and found strong absorption in the region 1210-1230 cm^{-1}, which is attributed to the C=S structure in the molecule. Little and Leja[2] obtained the spectra of alkali metal xanthates with different hydrocarbon chains, diethyl dixanthogen, and ethyl xanthates of copper, lead, and zinc. Figures 6-1 to 6-4 show the spectra of the different xanthate compounds. The frequencies of absorption

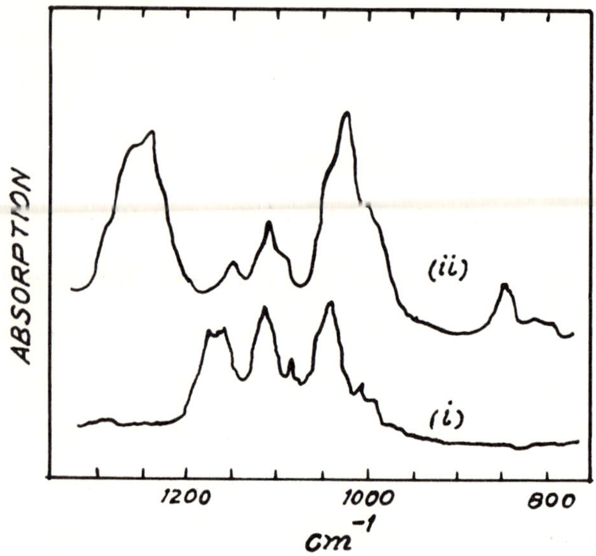

FIG. 6-1

Infrared spectra of xanthate compounds. (i) Sodium ethyl xanthate and (ii) diethyl dixanthogen. [After L.H. Little and J. Leja: Proceedings of the Second International Congress on Surface Activity, Vol. III, London, 1957, pp. 262-264.]

bands are recorded in Table 6-1. The results show that the precipitate of copper ethyl xanthate obtained on mixing cupric sulfate with potassium xanthate consists of a mixture of xanthate and dixanthogen because when the precipitate is extracted with ether the bands characteristic of dixanthogen disappear and the spectrum of the residue from the ether extract is indentical with that of dixanthogen. This is clear structural

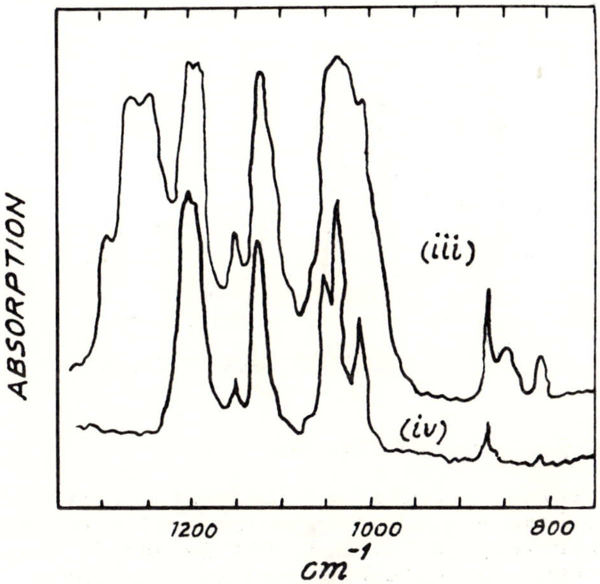

FIG. 6-2

Infrared spectra of xanthate compounds. (iii) Copper ethyl Xanthate and (iv) Cuprous ethyl xanthate. [After L.H. Little and J. Leja: Proceedings of the Second International Congress on Surface Activity, Vol.III, London, 1967, pp. 262-264]

evidence for the oxidation of xanthate to dixanthogen by cupric ions, which is in accord with the chemical evidence mentioned in Chapter 3. The authors also found that whereas the copper xanthate precipitated with cupric salt contains about equivalent amounts of dixanthogen and cuprous xanthate, the lead xanthate may contain about 5-10% dixanthogen (as estimated from the shoulder appearing at

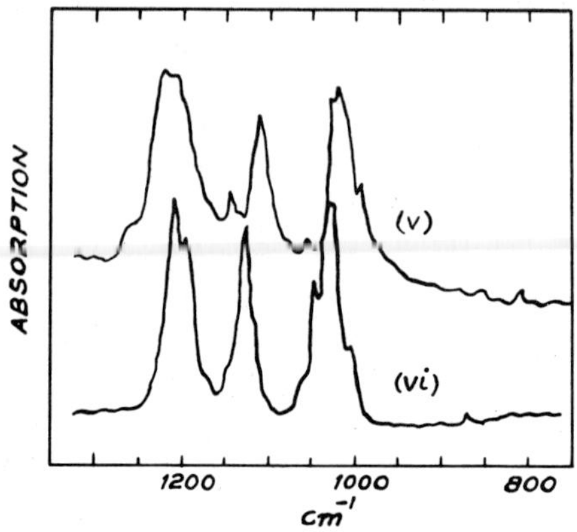

FIG. 6-3

Infrared spectra of xanthate compounds. (v) Lead ethyl xanthate and (vi) zinc ethyl xanthate. [After L.H. Little and J. Leja: Proceedings of the Second International Congress on Surface Activity, Vol. III, London, 1957, pp. 262-264.]

1260 cm^{-1}), while the spectrum of zinc xanthate shows no evidence of dixanthogen bands. It is not easy to suggest how dixanthogen arises in lead xanthate precipitate, since the reduction of lead to a lower oxidation state is not known.

The absorption bands in the region 1000-1200 cm^{-1} in the spectra of alakli metal xanthates are not well resolved, but it is possible to distinguish three more-or-less distinct groups of bands corresponding to the bands at 1200, 1130, and 1060 cm^{-1} for the heavy metal xanthates.

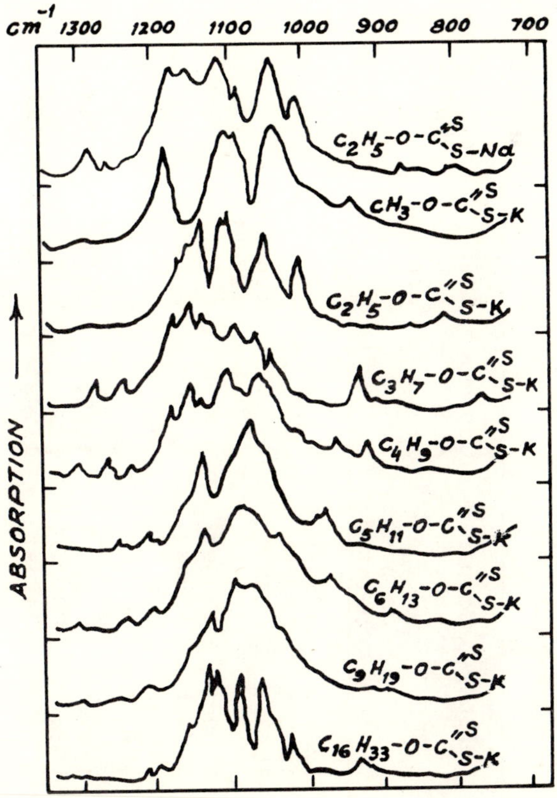

FIG. 6-4

Infrared spectra of alkali metal xanthates of different hydrocarbon chains. [After L.H. Little G.W. Poling and J. Leja: Can. J. Chem., 39, 745 (1961); reproduced by permission of the National Research Council of Canada.]

A detailed analysis of the infrared spectra has been described by Shankaranarayana and Patel.[3] They assigned values for the characteristic frequencies of C=S and C–O groups and compared them

TABLE 6-1

Frequencies of Adsorption Bands (cm^{-1}) of Xanthate Compounds[a,b]

Pure NaEtX	Pure KBuX	Pure KLX	$(EtX)_2$	Copper ethyl xanthate as precipitated	Cuprous ethyl xanthate	Lead ethyl xanthate	Zinc ethyl xanthate
—	1305w	—	1290sh	1290sh	—	—	—
	1262w		1262st	1265st			
			1242st	1242st		1260sh	
	1231w					1220st[c]	(1212st[c]
		1200w	1200st	1200st[c]			(1196st
			1190sh	1190sh			
1179st	1175m						
1160st		1162w					
	1153st		1152w	1152w	1152w	1145w	1148w
	1137m	1138st					
		1123st		1124st	1124st[c]		1127st[c]
1115st		1113sh				1112st[c]	
	1107st		1110m	1110sh			
1085w			1092sh	1094sh			
	1077st	1075st					
	1065sh	1061w		1067sh	1067sh		1065sh

Table 6-1 (Con't)

				1050sh	(1050sh[c]	1055w	(1049sh[c]
1040st			1040sh	1038st	(1037st		(1035st[c]
		1030w	1024st	1024st	(	(1026sh[c]	(1027sh
	1018w				(	(1020st	(
1010w			1000sh	1012sh	(1012m		(1007sh
			1000sh	1000sh		995sh	
	963w						
	926w	937w					
				872w	872w		872w
			849m	949w		853w	
			812w	813w	812w	806w	

[a] After L.H. Little and J. Leja: Proceedings of the Second International Congress on Surface Activity, Vol. III, ed. J.H. Schulman. Butterworth, London, 1957, p. 263.

[b] W, Weak; m, medium; St, Strong; Sh, Shoulder.

[c] Characteristic bands of heavy metal xanthates.

for various xanthate compounds. These values are listed in Table 6-2.

The xanthate ion can be represented in the following three forms:

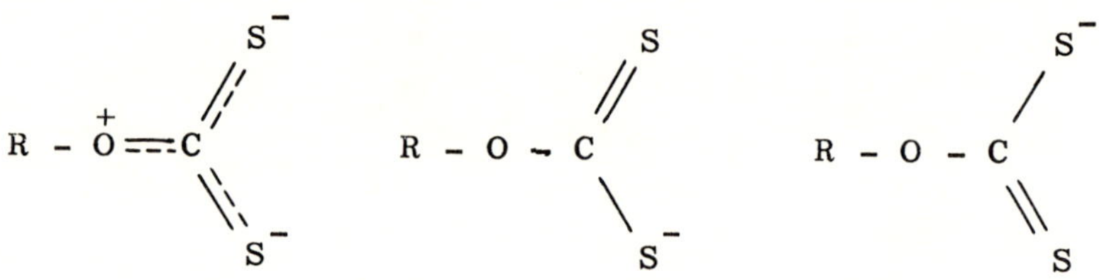

I II III

In ionic compounds the structure can be represented as a canonical structure (I). In such a case the π-bond character of C=S will be lowered, while the σ -bond of C–O will acquire a partial π-bond character. This would result in a lowering of the C–S frequency and an enhancement of the C–O frequency. This is shown to hold true, as revealed in the values entered in Table 6-2. The C=S frequency in sodium xanthate is lowered to 1179 cm^{-1}, while that of C–O is raised to 1040 cm^{-1}, as compared to the respective frequencies of the dixanthogen (nonionic molecule) at 1258 and 1026 cm^{-1}. If sodium is replaced by potassium, the ionic character of the xanthate ion will increase because of the lower electronegativity of the potassium ion. The

TABLE 6-2

Characteristic Frequencies of C=S and C-O Groups of the Derivatives of Xanthic Acid[a]

Substance	Medium	C=S Frequency, cm^{-1}	C-O Frequency, cm^{-1}
Sodium ethyl xanthate	Nujol mull	1179	1040
Potassium ethyl xanthate	KBr Disk	1143	1055
Potassium n-butyl xanthate	Nujol mull	1153	1077
Cuprous ethyl xanthate	Nujol mull	1200	1037
Zinc ethyl xanthate	Nujol mull	1212	1035
Lead ethyl xanthate	Nujol mull	1220	1020
Diethyl dixanthogen	Nujol mull	1242) (Split 1262	1026
Diisopropyl dixanthogen	CCl_4 solution	1258	1010
Di-n-butyl dixanthogen	CCl_4 solution	1258	1026
Diisoamyl dixanthogen	CCl_4 solution	1258	1024
Ethyl xanthic S-ethyl ester	–	1190	1053-1000 (Triply split band)

[a] After M.L. Shankaranarayana and C.C. Patel: Can. J. Chem., 39, 1634 (1961); reproduced by permission of the National Research Council of Canada.

π-bond character of C=S decreases further, while that of C-O increases resulting in further lowering of the C=S frequency and enhancement of the C-O frequency. This is also revealed in Table 6-2. The frequency of C=S in the potassium salt is lowered to 1143 cm^{-1}, while that of C-O is raised to 1055 cm^{-1}. When the ethyl group is replaced by an n-butyl group, both the C=S and C-O bonds move to higher frequencies, probably because of the greater electron-releasing tendency of the n-butyl group.

In the case of a covalent xanthate such as cuprous ethyl xanthate, the resonating structures II and III predominate. No bonding is expected between the copper atom and the π-bonded sulfur The π-bond character of C=S and the σ-bond character of C-O are greatly enhanced; this gives rise to a shift of C=S to the higher frequency and CO to the lower one. This has been confirmed experimentally. It is thus seen that as the metal-sulfur bond strengthens, the CO vibration frequency and its bond order are lowered. The relative shifts of C=S and C-O frequencies of zinc and lead ethyl xanthates are, however, not in accord with the electronegativities of their metal atoms. It is possible that other considerations such as the thermodynamic stability and the stereochemistry of the metal xanthates may also be contributory factors in determining the shifts of the C=S and C-O frequencies in such compounds.

The absorption spectra of four dixanthogens are shown in Fig.6-5. The spectra do not show any appreciable change in the C-S and C-O frequencies even though the alkyl groups of different dixanthogens vary. A small shift of the C-O frequency to 1010 cm^{-1} in the case of diisopropyl dixanthogen is probably attributable to the vibrational coupling of the C-O group with the C-C skeletal stretching vibration. Diisoamyl dixanthogen also has a similar branched hydrocarbon chain, but its remoteness from the C-C group prevents the interaction, and there is no shift in the C-O frequency. The constancy of the C=S group frequency at 1258 cm^{-1} in the dixanthogens is attributable to the symmetry of the groups in all parts of the molecule.

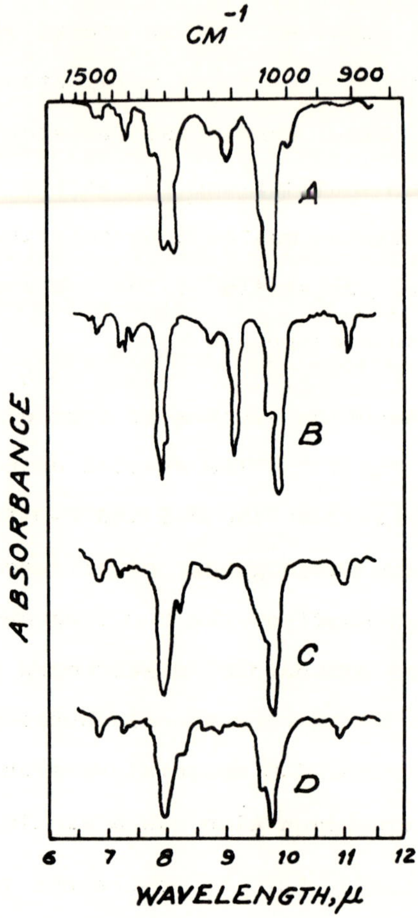

FIG. 6-5

Infrared spectra of dialkyl dixanthoges. (A) Diethyl dixanthogen, (B) diisopropyl dixanthogen, (C) di-n-butyl dixanthogen, (D) diisoamyl dixanthogen. [After M.L. Shankaranarayana and C.C. Patel, Can. J. Chem., 39, 1636 (1961); reproduced by permission of the National Research Council of Canada.]

The assignment of the C=S frequency around 1200 cm^{-1} and the CO frequency around 1030 cm^{-1} as indicated in the studies of Shankaranarayana and Patel[3] differs slightly from the values assigned by Little and co-workers[4] based on effect of organic solvents. These authors measured the band displacements of the 1020-1070 cm^{-1} band and the 1200-1280 cm^{-1} band in various solvents and represented the values relative to the frequency in cyclohexane. Table 6-3 shows their results. The frequency shifts ($\Delta\nu$) of the 1020-1070 cm^{-1} bands are all positive and of the order of 0-9 cm^{-1}, while those of the 1200-1280 cm^{-1} band are toward higher frequencies (negative) in the same solvent sequence. This evidence indicates that the 1020-1070 cm^{-1} band and not the 1200-1280 cm^{-1} band follows the C=S behavior, since positive shifts have been shown for other compounds with C=S groups, such as thiocarboxylic acids, investigated by other workers. According to Leja and co-workers, the resonance hybrid 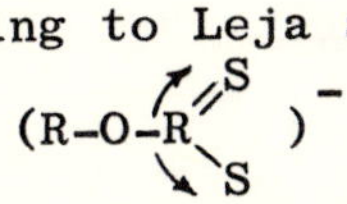 may be considered to represent the structure of the alkali metal xanthates. The structure of the heavy metal xanthates is, however, represented by 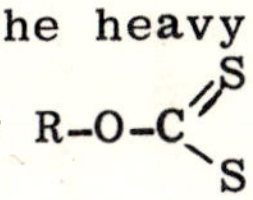 because they have the same

TABLE 6-3

Absorption Band Frequencies and Frenquency Shifts[a]

	1020-1070 cm^{-1} Band				1200-1280 cm^{-1} Band			
	Zn n-Butyl xanthate		n-Butyl dixanthogen		Zn-n-Butyl xanthate		n-Butyl dixanthogen	
	ν	$\Delta\nu$	ν	$\Delta\nu$	ν	$\Delta\nu$	ν	$\Delta\nu$
Cyclohexane	1046	0	1028	0	1238	0	1264	0
Diethyl ether	~1050	–	~1030	–	1203	35	1266	-2
Carbon tetrachloride	1044	2	1027	1	1244	-6	1266	-2
Benzene	1062) 1030)	-	1023	5	1241	-3	1268	-4
Dioxane	Solvent absorption		1022	6	1208	30	1269	-5
Acetonitrile	1060) 1037)	-	1021	7	1235) 1200) 1180)	-	1273	-9

Table 6-3 (Con't)

Nitromethane	Insoluble		1023	5	Insoluble		1274	-10
Methylene chloride	1039	7	1024	4	1242	-4	1280) 1248)	-
Chloroform	1040	6	1026	2	1245	-7	1273	-9
Bromoform	1038	8	1020	8	1247	-9	1267	-3
1,1,2,2-Tetrabromoethane	1037	9	1020	8	1255) 1213)	-	1267	-3
Solid or liquid	1045	1	1022 liq.	6	1209	29	1264 liq.	0

[a] After L.H. Little, G.W. Poling and J. Leja: Can. J. Chem., 39, 1783 (1961); reproduced by permission of the National Research Council of Canada.

C=S band at 1060 cm^{-1} and do not exist as resonance hybrids. For similar reasons the structure of dixanthogens retaining the C=S group is represented by

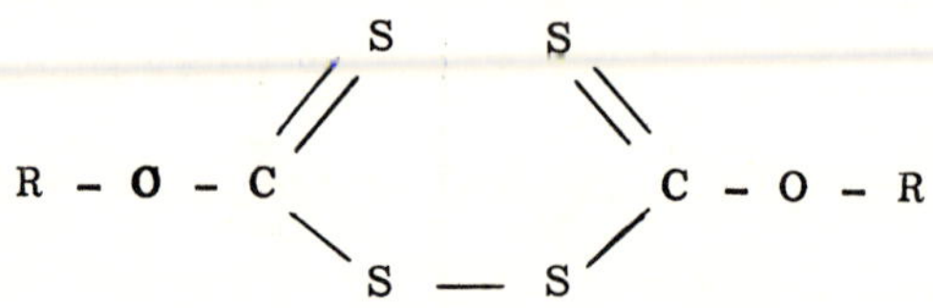

The assignment of C=S strectching frequency in the region 1200 cm^{-1} is to some extent supported by the spectral studies on infrared organosulfur compounds by Rao and associates.[5] According to these authors, potassium alkyl xanthates should be the limiting case of the dependence of the C=S stretching frequency on the double-bond chracter of the thiocarbonyl linkage. Therefore, asymmetric and symmetric CS_2^- stretching frequencies should be expected. If a medium intensity band in the range 1060-1045 cm^{-1} is tentatively assigned for the asymmetric stretching vibration and a weak band between 833-819 cm^{-1} to the symmetric stretching vibration, it is found that the asymmetric stretching bands show splitting in the solid state and that the C-O stretching frequency in potassium alkyl xanthates is considerably reduced (1186-1140 cm^{-1})

compared to that in dialkyl dixanthogens (1260 cm^{-1}). These results are not to be expected on the basis of the above assignment, since an increase in C-O frequency would be expected because of the different canonical structures (see I, II, and III, above) of the xanthate ion. This indicates that the assignments of C-O and C=S stretching frequencies in dialkyl dixanthogens should be reversed and that the low frequency assignment (1045-1010 cm^{-1}) to C-O and the high frequency assignment to C=S as reported by Shankaranarayana and Patel[3] should be accepted. However, Rao et al.[5] have mentioned the difficulty in accepting such a low frequency for C-O stretching in structures of xanthate compounds in which "contiguous" C-O bands would be expected in the higher frequency range.

Supporting evidence in favor of assigning 1030 cm^{-1} to the C=S group has been obtained[6] from the infrared spectra of the derivatives of trithiocarbonic acid. In these compounds the C-O-C group is replaced by a C-S-C group. The spectra of these compounds show a strong band in the region 1015-1050 cm^{-1} and no band is observed in the region of 1200 cm^{-1}, indicating thereby that the band at

1200 cm^{-1} observed in the spectra of xanthate compounds cannot be attributable to C=S group.

Dialkyl dixanthogens show C–S stretching between 680 and 605 cm^{-1} and S–S stretching between 510 and 505 cm^{-1}. The C–O and C=S stretching bands show considerable splitting which may be the result of Fermi interactions. The combination tone of C–S and S–S stretching vibrations seems to interact with the C–O stretching vibration (1270-1255 cm^{-1}), while the overtone of the S–S stretching frequency seems to interact with the C=S stretching vibration (1045-1010 cm^{-1}).

The difficulty in making an unambiguous assignment for C=S and C–O–C vibrations has recently been pointed out by Shankaranarayana and Patel[6] and Watt and McCormick.[7] This difficulty arises from the high coupling of the C=S and C–O motions which are also sensitive to environmental changes, especially in the alkali metal xanthates. Greater evidence, however, is in favor of assigning 1030 cm^{-1} to the C=S group. The two bands, one round 1030 cm^{-1} and the other around 1200 cm^{-1} may be considered to characterize the C–O–C(=S) group.

The assignment of other bands observed in the spectra of xanthates and dixanthogens has been made by Shankaranarayana and Patel.[6] The band occurring around 1140 cm^{-1} in the heavy metal alkyl xanthates is intense, while the corresponding band in dixanthogens is weak. This band is not found in the derivatives of trithiocarbonic acid, indicating that the band is associated with $-O-\overset{\|}{C}-$ group vibration. The intense absorptin band at 1087 cm^{-1} in diisopropyl dixanthogen may be assigned to the rocking mode of vibration of methyl groups coupled strongly with the skeletal vibration of an isopropyl group. The medium band around 900 cm^{-1} in the derivatives of xanthac acid is attributable to the $-\overset{\|}{C}-S$ group.

Watt and McCormick[7] have made a detailed study of the infrared spectra of methyl and ethyl xanthate complexes of transition metals, namely, nickel, chromium, cobalt, palladium, and platinum. Typical spectra are shown in Figs. 6-6 and 6-7, and spectral data are recorded in Tables 6-4 and 6-5. Assignments for the various complexes were inferred from spectral comparisons with bis(ethyl xanthate) nickel(II). The 63 normal modes of

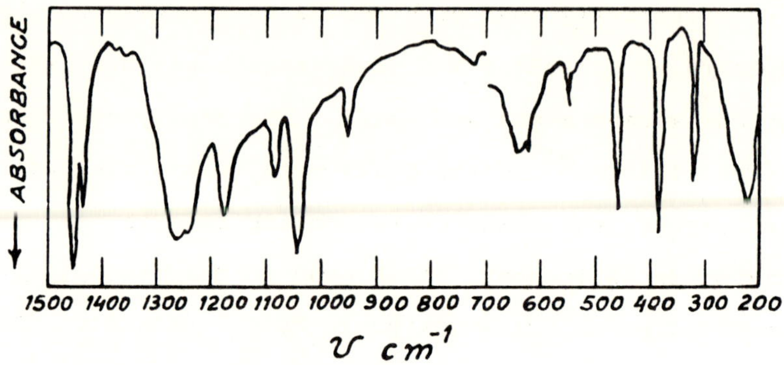

FIG. 6-6

Infrared spectrum of a methyl xanthate complex of nickel. [After G.W. Watt and B.J. McCormick: Spectrochim. Acta, 21, 754 (1965).]

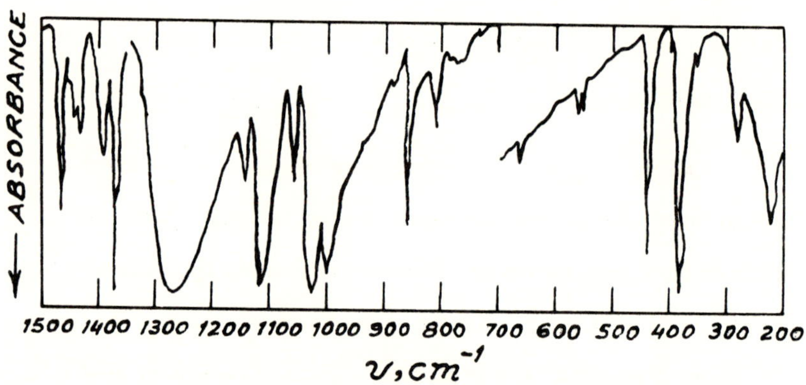

FIG. 6-7

Infrared spectrum of an ethyl xanthato complex of nickel. [After G.W. Watt and B.J. McCormick: Spectrochim. Acta, 21, 754 (1965).]

TABLE 6-4

Vibrational Assignments for Absorption Bands (cm^{-1}) in the Infrared Spectra of Complexes

$Cr(etxn)_3$	$Co(extn)_3$	$Ni(etxn)_2$	$Pd(etxn)_2$	$Pt(etxn)_2$	Assignment
2995m	2989m	2990m	2998m	2998m	—
2975m	2975m	—	2965m	2988m	$\nu(CH_3)$ and
2962vw	2960w	- -	—	—	and
2930w	2925w	-	2945w	2945w	$\nu(CH_2)$
2900vw	2895vw	2895vw	2900vw	2900vw	
2470w	2490w	2510vw	2510w	-	
1890v,w	1860b,vw	1855vw	1850vw	-	combination
1719w	1723w	1711w	1709w	1700b,w	
1390w	1390m	1390m	1390m	1389w	$\delta(CH_2)$
1369vs	1369vs	1369vs	1370vs	1771vs	
1453w	1453w	1449sh,w	1448sh,w	1451vw	
1446m	1446m	1444m	1444m	1447vw	$\delta(CH_3)$
1435m	1435m	1432m	1436m	1437vw	
1465m	1465m	1466m	1465m	1465w	

Table 6-4 (Con't)

1300sh,w	1300sh,w	1325w	1320w	1320sh,w	
1275vs	1276vs	-	1277s	-	
1250vs	1276vs	1268vs	1265s	1285vs	ν(C-O)
1120m	1134m	1115vs	1114s	1114s	
1144w	1144w	1144w	1143w	1141w	ω (CH_2)
1055w	1056w	1059w	1058vw	1058vw	τ (CH_2)
1032s	1035m	1025vs	1020s	1017vw	ρ (CH_3)
1004s	1005m	998vs	998m	1004m	
935b,vw	-	-	-	-	
-	-	880vw	-	-	
861m	865m	857m	856m	853m	ν (C-C)
814w	813	809w	810vw	810vw	ρ (CH_2)
760vw	-	-	-	-	
651w	658w	663w	660w	657w	
563vw	549vw	560vw	569vw	549vw	ν (C-S)
-	-	555vw	555vw	545vw	
-	618w	-	-	-	
442m	441m	437s	438s	442m	δ (C-O-C)
398s	393m	395vw	394m	394w	δ (C-O-C)
361vs	358vs	383vs	340vs	338vs	
340sh,w	340m	351vw	323m	312vw	ν (M-S)
321w	308m	-	-	-	
296vw	-	283w	-	-	δ (O-C-S)

a

b After G.W. Watt and B.J. McCormick, Spectrochim. Acta, 21, 756-757 (1965)

vw, Very weak; w, weak; m, moderate; vs, very strong; s, strong.

TABLE 6-5

Vibrational Assignments for Absorption Bands (cm^{-1}) in the Infrared Spectra of $M(mexn)_{2,3}$ Complexes[a,b]

$Cr(mexn)_3$	$Co(mexn)_3$	$Ni(mexn)_2$	$Pd(mexn)_2$	$Pt(mexn)_2$	Assignment
3330w	3035vw	3030w	3035w	3035w	—
3000vw	-	3000w	3000w	2995w	$\nu(CH_3)$
2950m	2953w	2945m	2945m	2950m	—
2940m	2940m	-	-	-	
2840w	2850w	2845vw	2840vw	2845vw	
1875b,w	1885w	1885b,w	1885b,w	1886b,w	combination (?)
1250 + 619	1250 + 625	1243 + 637	1277 + 612	(?)	
1455vs	1454s	1455vs	1456vs	1465vs	
1448vs	-	-	-	1445m	$\delta(CH_3)$
1440vs	1441s	1435s	1434s	1431s	
1440vs	1260b,vs	1263b,s	1277b,vs	1325b,s	
-	-	1243b,s	-	-	$\nu(C{-}O)$
1173m	1176m	1176m	1184m	1188w	
1123vw	-	-	-	-	Impurity band
1084v	1089m	1085w	1076w	1066w	
619 + 459	625 + 455	624 + 458	616 + 457	608 + 455	combination (?)

Table 6-5 (Con't)

1045sh,s	1042vs	1042s	1034vs	1024s	
1038s	1030sh,s	-	-	-	ρ (CH_3)
-	1010sh,w	-	1005sh,w	1014sh,s	
944w	946m	954w	947w	949w	
790w	-	-	-	-	Impurity bands
768vw	-	-	-	-	
-	675vw	-	-	-	
619vw	625vw	637m	616vw	608vw	
-	-	624m	612vw	-	ν (C-S)
563vw	548vw	548vw	552vw	553s	
-	-	-	544m	-	
459s	455vs	458s	458vs	455vs	δ (C-O-C)
376vs	361vs	383vs	347vs	362vs	
338vw	342m	260vs	335s	330m	ν (M-S)
324vw	336m	-	-	-	
318vw	-	-	-	-	
290vw	278vw	320s	305m	289s	δ (S-C-O)

[a] After G.W. Watt and B.J. McCormick: Spectrochim. Acta, 21, 756-757 (1965)

[b] Symbols have same meaning as in Table 6-4.

vibration for the nickel complex are distributed on the basis of C_{2h} symmetry among the various groups of the ligand and the S_2-Ni-S_2 skeleton. These vibrational relationships are shown in Table 6-6. The table also includes the vibrational mode distribution for nickel methyl xanthate complex derived on the assumption of C_{2h} molecular symmetry and Ci site symmetry.

All 10 complexes exhibit bands in the 2900-3100 cm^{-1} region. These are assigned to C-H stretching modes. Under the symmetry prevailing in nickel ethyl xanthate complex, there should be one infrared active CH_2 bending deformation mode, and for this reason the single absorption at 1390 cm^{-1} is assigned. The sharp absorptions and the shoulder at 1369-1432, 1444, 1466, and 1449 cm^{-1}, respectively, are assigned to CH_3 deformations; although only three methyl deformations are expected, the additional absorptions could be attributed to the weak vibrations in the local crystal field environment. In the methyl xanthates the methyl deformations are assigned at 1455 and 1435 cm^{-1}. The weak to moderately intense absorptions at 1144, 1059, and 809 cm^{-1} are assigned to CH_2 wagging, twisting, and rocking modes, respectively. In nickel xanthate

TABLE 6-6

Distribution of Vibrational Modes for $Ni(etxn)_2$ and $Ni(mexn)_2$ [a]

Mode	Representation
ν (CH_3)	$2A_g + B_g + A_u + 2B_u$
δ (CH_3)	$2A_g + B_g + A_u + 2B_u$
ρ (CH_3)	$A_g + B_g + A_u + B_u$
ν (CH_3) [b]	$B_g + A_u$
δ (CH_2) [b]	$A_g + B_g + A_u + B_u$
ρ (CH_2) [b]	$A_g + B_u$
ρ (CH_2) [b]	$B_g + A_u$
ν (C-C) [b]	$A_g + B_u$
ν (C-O)	$2A_g + 2B_u$
ν (C-S)	$A_g + B_g + A_u + B_u$
δ (C-C-O) [b]	$A_g + B_u$
δ (C-O-C)	$A_g + B_u$
π $(S_2$-C-O) [c]	$A_g + B_u$
δ (S-C-S)	$A_g + B_u$
δ (O-C-S)	$2B_g + 2A_u$
γ (O-C)	$B_g + A_u$
ν (Ni-S)	$A_g + B_g + A_u + B_u$

Table 6-6 (Con't)

δ (S-Ni-S)	$A_g + A_u + B_u$
π (S-Ni-S)	$A_u + B_u$

[a] vib $\left[Ni(etxn)_2\right] = 18A_g + 12B_g + 14A_u + 19B_u$;

vib $\left[Ni(mexn)_2\right] = 13A_g + 8B_g + 10A_u + 14B_u$

[b] Modes not applicable to $Ni(mexn)_2$

[c] Bend of S_2-C-O group out of S_2-Ni-S_2 plane.

the strong band at 1042 cm^{-1} is unambigously assigned to methyl rocking motion. The weak-to-medium band at 954 cm^{-1} is tentatively assigned as the remaining methyl rocking mode. The C-C vibration is assigned at 857 cm^{-1} in the ethyl xanthates, since a corresponding absorption is not found in the spectra of the the methyl xanthates. Two bands, one of which is sometimes split, occur in the region 540-670 cm^{-1} and vary greatly in intensity. They are assigned as C-S stretching frequencies.

The C-O-C deformation in nickel xanthate complex can be assigned at 437 cm^{-1} since a band similar in intensity, shape, and frequency occurs in all xanthates including those of potassium. This C-O-C frequency lies somewhat higher in the methyl xanthates because of the mass effect.

The weak absorption band at 283 cm^{-1} in nickel ethyl xanthate complex and the one at 320 cm^{-1} in nickel methyl xanthate complex are tentatively assigned to the O-C-S deformation mode, and corresponding absorptions are so assigned in the remaining methyl xanthate complexes.

The theoretical analysis as shown in Table 6-5 predicts that there are two infrared-active Ni-S stretching frequencies in nickel ethyl xanthate complex. One of these can be assigned with certainty to the very strong and sharp absorption at 383 cm^{-1}. A corresponding band occurs at 383 cm^{-1} in methyl xanthato complex. The assignment of the remaining Ni-S stretching is somewhat difficult. However, the weak absorption at 351 cm^{-1} is tentatively assigned as the second Ni-S stretching frequency. This assignment is made in view of the spectra of ethyl xanthato complexes of cobalt and palladium and methyl xanthato complexes of cobalt, palladium, and platinum, all of which show a definite, moderately strong band at 20-30 cm^{-1} below the bands of nickel xanthate, that is 358, 340, 361, 347, and 262 cm^{-1}, respectively. Nevertheless the assignments of the second M-S stretching frequency in these complexes must be regarded as tentative.

The intense band in the range 1280-1325 cm^{-1} in xanthate complexes is primarily attributable to a C-O vibration and if this is considered indicative of a C-O-C group, the resonance structure of the complex is expressed as

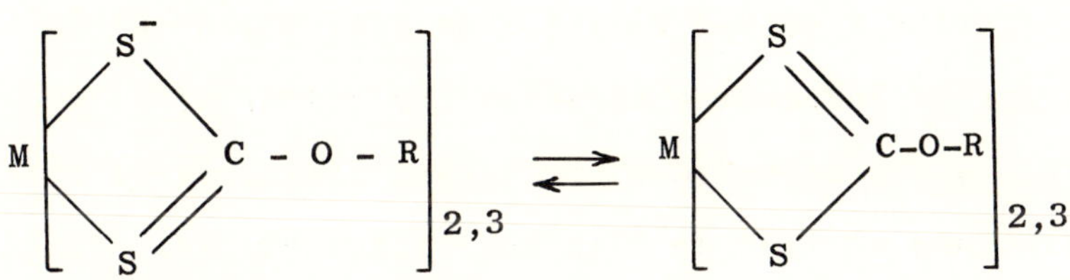

An alternate structure,

a

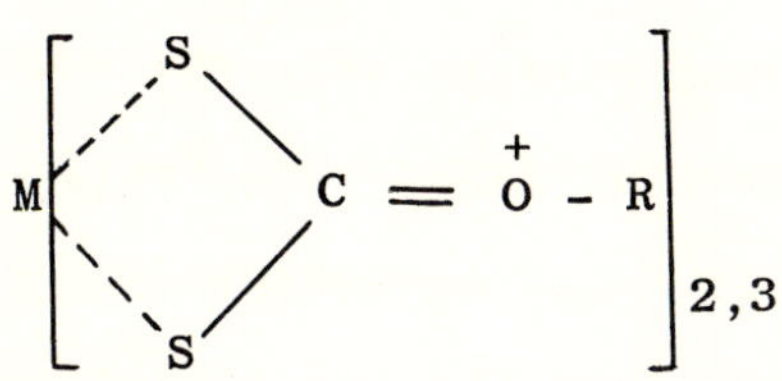

b

cannot be ruled out because the assignment of the

frequency to a C-O-C linkage is rather ambiguous, as mentioned in the works of Little et al.[4] and Shankaranarayana and Patel.[3,6] It appears that "normal" C-O vibrational frequencies might be found anywhere between 1000 and 1250 cm^{-1} in xanthate compounds. Hence, there is no justification in rejecting structure b. Furthermore, information available from X-ray analysis (see below) suggests that structure b might make an appreciable contribution to the overall structure.

The position of the C-O band (in the range 1250-1325 cm^{-1}) increases in the order Cr(III) < Co(III) < Ni(II) < Pd(II) < Pt(II), which is the order of increasing affinity for sulfur as a donor atom. The second C-O stretching frequency is assigned at 1115 cm^{-1} in the nickel complex, thus there is a difference of about 153 cm^{-1} between the two C-O stretching frequencies in this complex. A similar band is found in the Cr(III), Co(III), Pd(II), and Pt(II) complexes at 1120, 1134, 1114, and 1114 cm^{-1}, respectively. In the methyl xanthate complexes the difference between the two frequencies amounts to 90-100 cm^{-1}, except in the case of the platinum complex which appears anomalous.

The observed shifts and splittings in C-O stretching frequencies indeed suggest that structure b is not insignificant in all xanthate complexes and that its relative contribution to the overall structure is determined largely by the nature of the metal involved. The importance of structure b cannot at present be expressed in quantitative terms. The available evidence indicates, however, that the relative contribution of structure b is considerably less than that of structure a .

ELECTRONIC ABSORPTION SPECTRA OF XANTHATES AND DIXANTHOGENS

The spectra of xanthates, dixanthogens, and xanthogenmonosulfide in the visible and ultraviolet regions in different solvents have been studied by Shankaranarayana and Patel.[8,9] They are summarized in Table 6-7; the spectra of potassium butyl xanthate in ethanol and dibutyl dixanthogen in heptane and in ethanol are shown in Figs. 6-8 and 6-9.

The xanthates studied exhibit a low-intensity absorption band around wavelength 380 mμ , and the corresponding band for dixanthogens occurs around

TABLE 6-7

Peak Wavelengths and Molar Extinction Coefficients for Xanthates and Dixanthogens[a]

Compound	Heptane		Ethanol		Acetonitrile		Water	
	λ_{max} mμ	log ϵ	λ_{max} mμ	log ϵ	λ_{max} mμ	log ϵ	λ_{max} mμ	log ϵ
Potassium methyl xanthate	–	–	228 304 382.5	3.97 4.31 1.71	–	–	226 302 381.5	3.90 4.28 1.71
Potassium butyl xanthate			229 304 386	4.00 4.30 1.80			229 301 381	3.90 - 1.82
Dimethyl dixanthogen	240 278 361	4.31 3.90 1.94	239 279 368	4.34 3.91 1.94	239 282 358	4.31 3.92 1.92		
Diethyl dixanthogen	240 279 363	4.34 3.91 1.90	368 281 361	4.32 3.90 1.91	239 283 360	4.34 3.91 1.93		

Table 6-7 (Con't)

Dibutyl dixanthogen	244 286 365	4.38 3.92 1.91	243 287 362	4.37 3.91 1.92	243 288 360	4.38 3.90 1.92
Diisoamyl dixanthogen	245 286 367	4.39 3.95 1.92	243 287 363	4.38 3.92 1.91	242 288 360	4.38 3.90 1.91
Dibutyl xanthogen monosulphide	225 285 365	4.30 3.90 1.90	224 286 363	4.34 3.80 1.91		
Diisopropyl dixanthogen	243 285 365	4.36 3.92 1.91	241 286 362	4.35 3.94 1.90	240 287 367	4.30 3.91 1.93

[a] After M.L. Shankaranarayana, and C.C.Patel, Can. J. Chem., 39, 2590-2592 (1961); Acta Chem. Scand., 19, 113-1119 (1965).

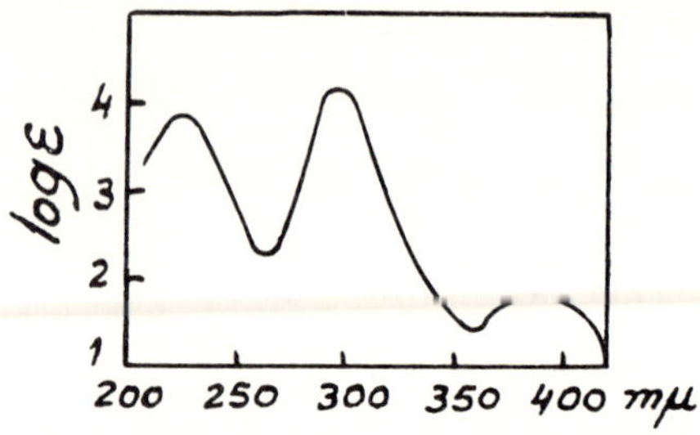

FIG. 6-8

Spectrum of potassium butyl xanthate in ethanol. [After M.L. Shankaranarayana and C.C. Patel: Acta Chem. Scand., 19, 1116 (1965).]

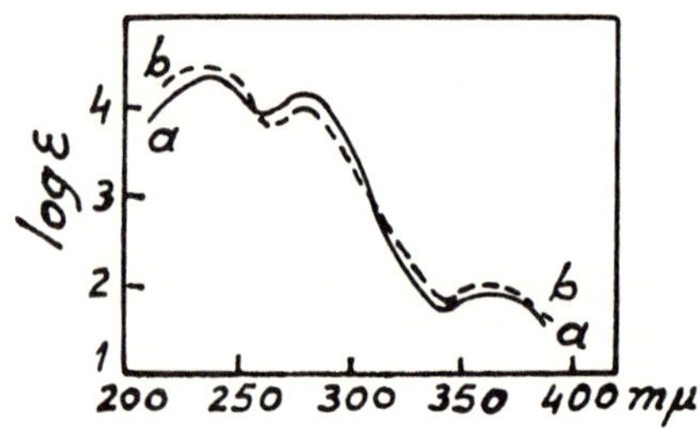

FIG. 6-9

Spectra of dibutyl dixanthogen in heptane (a) and in ethanol (b). [After M.L. Shankaranarayana and C.C. Patel: Acta Chem. Scand., 19, 1116 (1965).]

360 mμ . This band is very sensitive to the solvent used and shows large blue shifts progressively as the solvent is changed from nonpolar to polar. This implies that the transition energy involved increases with an increase in the polarity of the solvent because of solvation and/or hydrogen

bonding between the nonbonding electrons of the thiocarbonyl group of the solute and the solvent molecules. The low-intensity band is ascribed to the excitation of a nonbonding electron localized on the sulfur atom of a thiocarbonyl group to an antibonding π^{*} orbital. This transition is known as an n-π^{*} transition. The blue shift observed when the solvent is changed from a nonpolar to a polar nonhydrogen-bonding one (acetonitrile) is explained by the formation of an electrostatic bond between the nonbonding electrons of the thiocarbonyl sulfur atom and the positively charged carbon atom of the nitrile group. Such a bond lowers the energy of the n-electron orbital and, upon excitation, the electrostatic bond is either broken or weakened.

The intense band (log $\epsilon \simeq 4$) around 285 mμ for dixanthogen shows a red shift when the solvent is changed from a nonpolar to a polar one. A small increase in red shift is observed with an increase in the polar character of the solvents. This band is ascribed to a $\pi - \pi^{*}$ transition in which an electron jump occurs from the highest occupied π orbital to the lowest empty (antibonding) π level.

The intense bands around 242 mμ (log $\epsilon \simeq 4.3$) for dixanthogen and around 300 mμ for xanthates are characterized by a blue shift as the polar character of the solvent is increased. This band is ascribed to a $n \rightarrow \sigma^*$ transition and is caused by the excitation of an electron from the nonbonding lone pair of electrons localized on the bonded sulfur atom in the S-S group to an antibonding orbital, either 3d or 4s. The shift in the band wavelength from dixanthogens to xanthates has not been explained. It has been observed that the wavelength of the $\pi \rightarrow \pi^*$ transition band decreases from 285 mμ to about 225 mμ and that of the $n \rightarrow \sigma^*$ band increases from 242 to 300 mμ with xanthates.

ELECTRIC MOMENTS OF XANTHATES

The earliest work on the structural characteristics of xanthates was reported by Malatesta,[10] who measured dielectric constants of several heavy metal xanthates in benzene and calculated their electric moments. There is a considerable difference between total polarizatin and electronic polarization of xanthates. This difference is assumed to correspond to an effective electric moment, which depends on the molecular structure. The electric moments of the metal xanthates are a little higher than the values calculated on the basis of a symmetric chelate structure, suggesting an imperfectly chelated structure for xanthates. This postulate is well supported by the physical and chemical characteristics and also by a recent study of the electron paramagnetic resonance (EPR) spectra of a series of copper xanthates.[11] The most probable structure assigned is

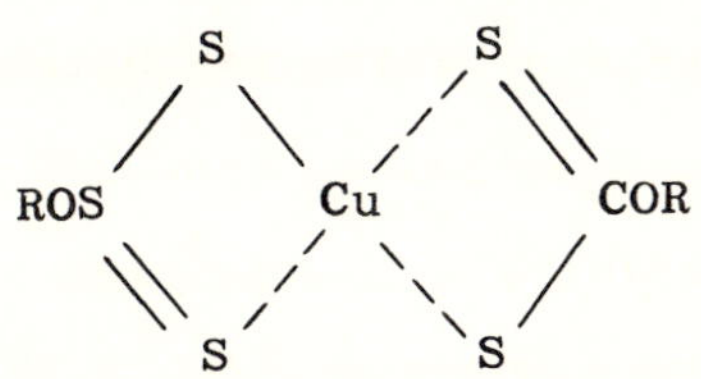

The bonds are mainly covalent in this structure.

CRYSTAL STRUCTURES OF XANTHATES

Although preliminary analyses of the crystal structure of a few xanthates were reported in 1938-1939 by a group of Italian workers, these are not easily accessible as they were published in Italian journals that were not too well known. The detailed structural analysis of several metal xanthates attracted much attention in the 1960's.

A major part of the work was done in Italy and Japan. The Japanese workers have also analyzed the crystal structure from the standpoint of collector action of xanthates on sulfide minerals and as such their data is of considerable interest to investigators in mineral flotation. The structural characteristics arrived at by X-ray analysis of the solid compounds closely agree with the results obtained by spectroscopic analysis, and it is now possible to describe the detailed structure of the xanthate group in different metallic xanthates.

POTASSIUM ETHYL XANTHATE

The crystal structure shown in Fig. 6-10 was determined by Mazzi and Tadini[12] by making (100) Weisenberg pictures as well as (010) and (001) Buerger photographs and refined by means of the consecutive applications of the following determinations: interpretation of a (100) Patterson projection (100) electron density projection; generalized projection along the same axis from the 1kl reflections and four consecutive applications of the least squares method to the structure factors of the 0kl, 3kl, h0l, hk0 reflections.

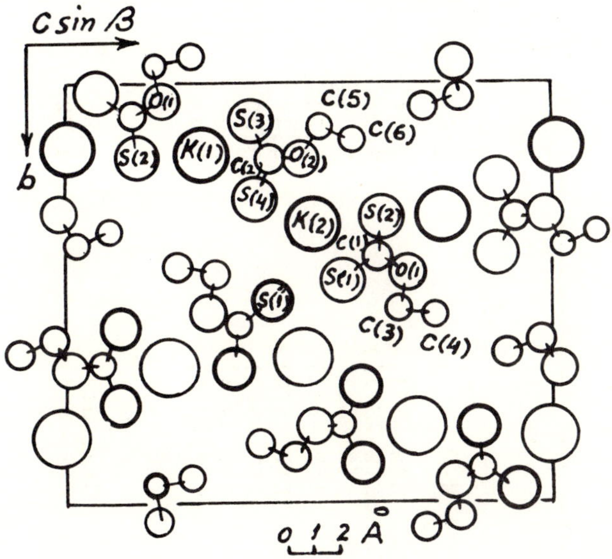

FIG. 6-10

The crystal structure of potassium xanthate viewed along (100). [After F. Mazzi and C. Tadini: Z. Krist., 118(5/6), 389, (1963).]

Each potassium atom is completely surrounded by sulfur atoms, but the coordination is slightly different for the two nonequivalent potassium atoms in the unit cell. While K(1) coordinates six sulfur atoms (see Table 6-8) S(2), S(3), and S(4) and their equivalents S(2'), S(3'), and S(4') of the adjacent cell roughly arranged at the corners

TABLE 6-8

Interatomic Distances and Bond Angles (see Fig. 6-10) in Potassium Ethyl Xanthate[a]

Atoms	Distance, Å	Atoms	Distance, Å
K-S	± 0.02	K-O	± 0.03
K(1)-S(2)	3.24	K(2)-S(1)	3.37
K(1)-S(2')	3.34	K(2)-S(1')	3.25
K(1)-S(3)	3.32	K(2)-S(1")	3.38
K(1)-S(3')	3.24	K(2)-S(2)	3.54
K(1)-S(4)	3.48	K(2)-S(2')	3.47
K(1)-S(4')	3.31	K(2)-S(4)	3.30
K(1)-O(1)	2.86	K(2)-S(4')	3.40
		K(2)-O(2)	2.48
C-S	± 0.05		
C(1)-S(1)	1.69	C(2)-S(3)	1.66
C(1)-S(2)	1.67	C(2)-S(4)	1.72
C-O	± 0.06		
O(1)-C(1)	1.35	O(2)-C(2)	1.34
O(1)-C(3)	1.48	O(2)-C(5)	1.45

Table 6-8 (Con't)

Bonds	Angle, degrees	Bonds	Angle, degrees
C–C ± 0.07 Å			
C(3)-C(4)	1.54	C(5)-C(6)	1.54
Bond angles in the xanthate molecule			
S(1)-C(1)-S(2)	124 ± 3	S(3)-C(2)-S(4)	125 ± 3
S(1)-C(1)-O(1)	121 ± 3.5	S(3)-C(2)-O(2)	122 ± 3.5
S(2)-C(1)-O(1)	114 ± 3.5	S(4)-C(2)-O(2)	113 ± 3.5
C(1)-O(1)-C(3)	116 ± 3.5	C(2)-O(2)-C(5)	121 ± 3.5
O(1)-C(3)-C(4)	103 ± 4	O(2)-C(5)-C(6)	107 ± 4

[a] After F. Mazzi and C. Tadini: Z. Krist., 118(5/6), 389 (1963).

of a distorted trigonal prism, K(2) has, in addition to a similar coordination, a further contact with an S(1''), which roughly centers one of the faces of the distorted trigonal prism. The average value of the bond distance K-S is 3.321 Å for K(1) and 3.387 for K(2); the increase in the second potassium atom is explained by the increase in the coordination number from 6 to 7. Each potassium atom is bound to one oxygen atom of the xanthate molecule; the K-O distances found (2.84 and 2.86 Å) are comparable to the corresponding bond distances reported in the literature.

The S-CS-O groups are planar, as expected for a carbon surrounded by three atoms. The C-O bond distance within this group shows that it has a partial double-bond character. Accordingly, the possibility of resonance among the following three structures is suggested for the xanthate group.

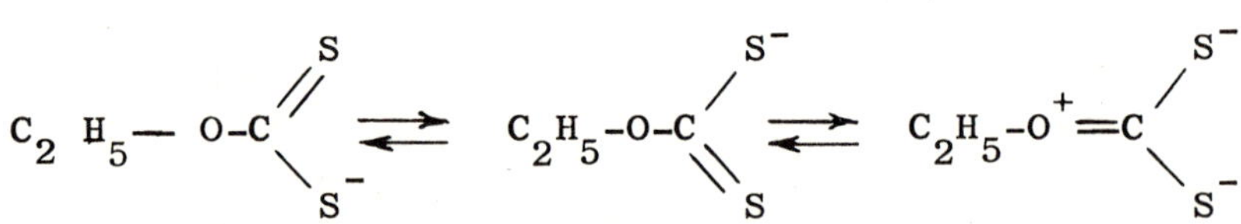

These forms are in accord with spectroscopic evidence discussed above.

The lattice parameters of potassium xanthate are as follows: a = 4.40 Å; b = 16.46 Å; c = 18.89 Å; $\beta = 92^\circ 7'$; space group, $P\,2_1/C$; Z = 8; D = 1.56. Interatomic distances and bond angles are shown in Table 6-8.

ARSENOUS XANTHATE

The earlier X-ray crystallographic data were checked by Carrai and Gottardi[13] by the use of Buerger photographs of arsenous xanthate. The space group assigned is $R\bar{3}$; a = 10.04 Å; $\alpha = 102^\circ 40'$. The structure was elucidated by the measurement of the diffraction intensities of the Buerger photographs. Three xanthate radicals are bound to each arsenic atom. Two molecules, $As(S\text{-}CS\text{-}O\text{-}C_2H_5)_3$, are superimposed and rotated 60°. Each aresenic atom is surrounded by three sulfur atoms denoted by S(1) that are bonded and by three nonbonded sulfur atoms denoted by S(2). The six sulfur atoms form a kind of distorted octahedron around arsenic. The interatomic distances and bond angles are shown in Table 6-9. Figure 6-11 represents a view of the atom assemblage in the xanthate radical.

TABLE 6—9

Interatomic Distances and Bond Angles in Arsenous Xanthate[a]

Atoms	Distance, Å	Atoms	Angle, degrees
As-S(1)	2.28 ± 0.007	S(1)-As-S(1)	92 ± 0.5
As....S(2)	2.94 ± 0.007	As-S(1)-C(1)	92 ± 1
S(1)-C(1)	1.66 ± 0.04	S(1)-C(1)-S(2)	129.5 ± 3
S(2)-C(1)	1.65 ± 0.04	S(1)-C(1)-O	112.5 ± 3
O-C(1)	1.31 ± 0.04	S(2)-C(1)-O	122.5 ± 3°
O-C(2)	1.45 ± 0.03	C(1)-O-C(2)	110 ± 4
C(2)-C(3)	1.49 ± 0.04	O-C(2)-C(3)	110 ± 4

[a] After G. Carrai and G. Gottardi, Z., Krist., 113, 373 (1960).

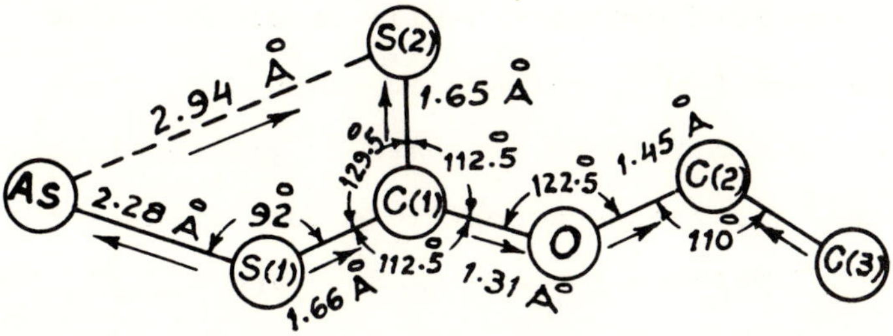

FIG. 6-11

Lower part of the arsenous xanthate molecule. The arrows along the bond lengths point toward the higher ones of the joined atoms.

The fact that the two distances C(1)-S(1) and C(1)- S(2) are very nearly equal suggests the possibility of resonance between the two forms:

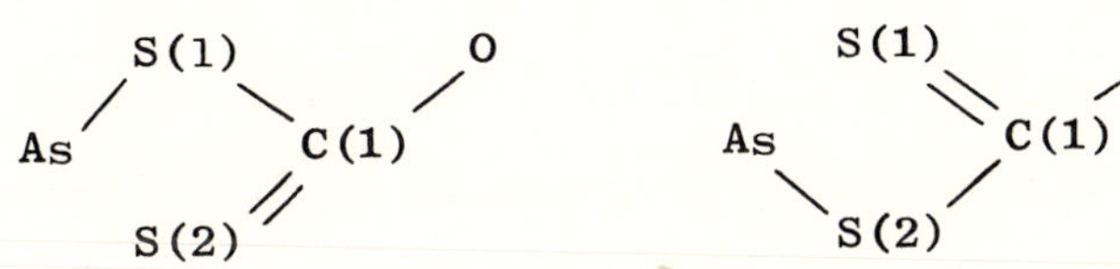

(I) (II)

The first form would be predominant. The distance As....S(2) is about 30% greater than the bond distance As-S(1). The attractive force between As....S(2) causes the increase in the As-S(1) distance and the closing of the S(1)-As-S angle. The S-CS-O group is planar. The C(1) atom is 0.055 Å from the S-CS-O plane.

The bond distance O-C(1) is close to the single bond C-O in carboxylic acids and esters, while the C-S bond in xanthate is greater than expected for a thiocarboxylic group because of the resonance between the two $As-S_1$ and $As...S_2$ bonds as noted above. In general the crystallographic data support the carboxylic structure for the xanthate radical.

ANTIMONOUS XANTHATE

The structure assigned to arsenous xanthate has been very closely confirmed for antimonous xanthate by the X-ray diffraction studies of Gottardi.[14] The space group assigned is $R\bar{3}$; $a = 10.13$ Å ; $\alpha = 103^\circ 30'$. The interatomic distances and bond angles are shown in Table 6-10.

The attraction between antimony atom and the unbound sulfur atom S(2) is greater than the corresponding attraction in arsenous xanthate and, consequently, the distance Sb...S(2) is only 20% greater than the distance Sb-S(1); in arsenous xanthate the corresponding difference is 30% of the smaller distance. The greater attraction is probably related to the greater amount of ionic character which is 12% in a Sb-S bond as compared to only 6%

TABLE 6-10

Interatomic Distances and Bond Angles in Antimonous Xanthate[a]

Atoms	Distance, Å	Bonds	Angle, degrees
Sb-S(1)	2.52 ± 0.006	S(1)-Sb-S(1)	97.5 ± 0.5
Sb....S(2)	3.00 ± 0.006	Sb-S(1)-C(1)	89 ± 1
S(1)-C(1)	1.70 ± 0.02	S(1)-C(1)-S(2)	130.5 ± 2
S(2)-C(1)	1.59 ± 0.02	S(1)-C(1)-O	108 ± 2
O-C(1)	1.36 ± 0.025	S(2)-C(1)-O	121 ± 2
O-C(2)	1.48 ± 0.025	C(1)-O-C(2)	120 ± 2
C(2)-C(3)	1.49 ± 0.04	O-C(2)-C(3)	110 ± 2

[a] After G. Gottardi, Z. Krist., 115, 451 (1961).

in an As-S bond. It may be mentioned here that in the case of potassium ethyl xanthate, in which the ionic character is very much greater, the two K-S bonds are identical and as a consequence the two C-S bonds C(1)-S(1) and C(1)-S(2) in Table 6-8 are practically equal to each other. While in the case of potassium xanthate the C-O bond also shows partial double-bond character in covalent xanthates, this bond is virtually a pure single bond.

The three bound sulfur atoms and the unbound ones form a distorted octahedron around the antimony atom. The mean distance Sb-S is 2.76 Å, while the sum of the ionic radii of Sb^{3+} and S^{2-} is 2.60 Å. The cation-anion radii ratio is 0.41, which is the exact ratio required for octahedral coordination. As a consequence of the attraction of $Sb...S_2$, the angle S(1)-Sb-S(1') is still smaller than the corresponding angle in arsenous xanthate and the same holds true for the Sb-S(1)-C(1) angle.

The geometry of the triangle S(1)-S(2)-O is the same in arsenous and antimonous xanthates. The group S(1)-CS(2)-O is confirmed to be planar, and the interatomic distances C(1)-O, C(2)-C(3), and

O-C(2) agree closely with the recent values reported in the literature for other organic compounds in which similar bonds occur (see Table 6-10).

NICKEL XANTHATE

The early work on the crystal structure of this compound by Bonatti[15] was repeated recently by Franzini,[16] who made some Weisenberg and some Buerger precession photographs using very small crystals. The space group was shown to be Pbca (= D_{2h}^{15}), with unit cell dimensions: a = 7.57 Å ; b = 7.23 Å. A schematic view of the nickel xanthate molecule is shown in Fig. 6-12. Each nickel atom is surrounded by four equidistant sulfur atoms; the nickel atom lies on a center of symmetry and, hence, the four sulfur and the nickel atoms are exactly on the same plane. Figure 6-12 shows the distribution in space of the covalent bonds attached to nickel. The equidistance of the bonds Ni-S(1) and Ni-S(2) is explained by a resonance structure of the type

As a result, the distance S(1)-C(1) should be equal to the distance S(2)-C(1) and, therefore, the

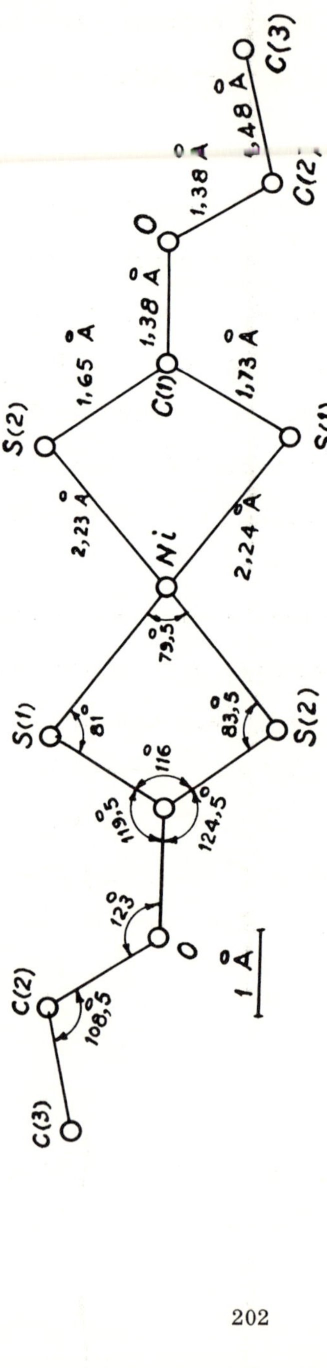

FIG. 6-12

Schematic view of the nickel xanthate molecule.
[After M. Franzini: Z. Krist., 118 (5/6), 393-403 (1963).]

two sulfur atoms are equivalent in respect to the C(1) atom. In actual fact, however, the distance S(1)-C(1) is 1.73 Å and the distance S(2)-C(1) is 1.65 Å; the latter bond has a higher double-bond character than the former one. This is explained by the interactions between different molecules in the structure, and is represented in Fig.6-13, which shows that the two sulfur atoms are no longer equivalent; the nickel atom nearest to S(1) is at a distance of 3.41 Å, while the atom nearest to S(2) is an oxygen atom at a distance of about 3.50 Å.

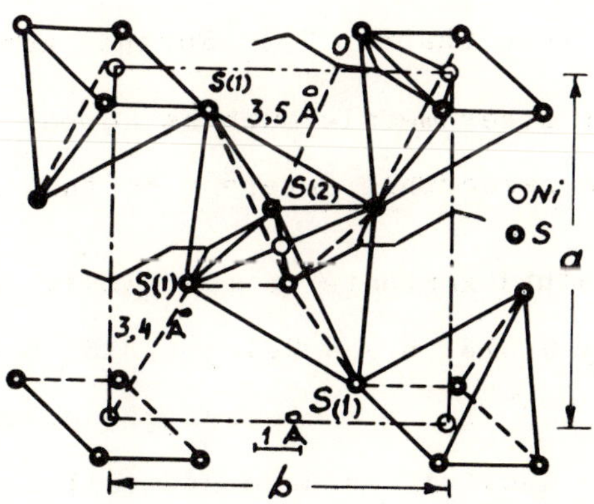

FIG. 6-13

Interactions between different molecules in the nickel-xanthate structure. [After M. Franzini: Z. Krist., 118 (5/6), 393-403 (1963).]

The nickel atom has four strong d-s- bonds directed to the corners of a square, and a fifth weak lp bond. This arrangement can result in a weak interaction with a sulfur atom belonging to another molecule. In the oxygen atom, all orbitals are occupied and, hence, the atom cannot interact with a sulfur atom of another molecule. The interatomic distances and bond angles calculated by Franzini are recorded in Table 6-11.

LEAD XANTHATES

Detailed crystal structural analysis has been recently described by Hagihara and co-workers,[17,18] and the data have provided useful knowledge in interpreting the selective flotation of galena (PbS) by xanthate (see Chapter 7). Further work on xanthates of other metals such as copper and iron will be of considerable interest in this area.

Lead ethyl xanthate is monoclinic, with: $a = 13.267 \pm 0.016$; $b = 4.301 \pm 0.015$, $c = 22.722 \pm 0.016$ Å; $\beta = 109^\circ\ 16' \pm 5'$; space group, $P2_1/c$; $Z = 4$. Lead n-butyl xanthate crystallizes into ordered and disordered structures. The disordered crystal is monoclinic, with: $a = 36.87 \pm 0.25$; $b = 4.58 \pm 0.11$; $c = 10.71 \pm 0.09$ Å; $\beta = 112.0 \pm 0.6^\circ$;

TABLE 6-11

Interatomic Distances and Bond Angles in Nickelous Ethyl Xanthate[a]

Atoms	Distance, Å	Bonds	Angle, degrees
Ni-S(1)	2.24 ± 0.006	S(1)-Ni-S(2)	79.5 ± 0.2
Ni-S(2)	2.23 ± 0.006	Ni-S(1)-C(1)	81 ± 0.5
S(1)-C(1)	1.73 ± 0.015	Ni-S(2)-C(1)	83.5 ± 0.5
S(2)-C(1)	1.65 ± 0.015	S(1)-C(1)-S(2)	116 ± 1
O-C(1)	1.38 ± 0.015	S(1)-C(1)-O	119.5 ± 1
O-C(2)	1.38 ± 0.015	S(2)-C(1)-O	124.5 ± 1
C(2)-C(3)	1.48 ± 0.020	C(1)-O-C(2)	123 ± 1.5
		O-C(2)-C(3)	108.5 ± 1.5

[a] After M. Franzini, Z. Krist., 118(5/6), 393 (1963).

space group, C2/c; Z = 4. The ordered crystal is also monoclinic. Its cell dimensions are the same as those of the disordered crystal except that the axis is quadruplicated. The interatomic distances and bond angles are given in Tables 6-12 and 6-13.

A molecule of lead ethyl xanthate forms a nonplanar asymmetric unit in the crystal. The deviations of carbon atoms from each place defined by two sulfur atoms and one oxygen atom are less than 0.15 Å. The C(3) atom deviates from the plane defined by S(1), S(2), and O(1) atoms by 0.38 Å. The lead atom deviates from the plane S(1)S(2)C(1) S(1)S(2)C(1)O(1)C(2) by 0.42 Å and from the plane S(3)S(4)C(4)O(2)C(5)C(6) by 0.79 Å. The two planes meet at an angle of nearly 117°. The asymmetry of the structure of the molecule about the lead atom is seen from the difference of 0.39 Å between the inter atomic distances S(2)-S(3) and S(1)-S(4).

The representation of the configuration in Figs. 6-14 and 6-15 shows the molecules arrayed in the crystal on twofold screw axes and symmetry centers. S(2) and S(2') atoms and S(4) and S(4') atoms come into contact through the twofold axes. Their distances (between 3.59 and 3.65 Å) are close

TABLE 6-12

Atom Distances, Bond Angles, and Their Standard Deviations in Lead Ethyl Xanthate[a]

Bond distances, Å	
Pb-S(1)	2.74 ± 0.03
C(1)-S(1)	1.70 ± 0.11
C(1)-S(2)	1.68 ± 0.11
C(1)-O(1)	1.30 ± 0.12
O(1)-C(2)	1.52 ± 0.14
C(2)-C(3)	1.53 ± 0.18
Pb-S(3)	2.79 ± 0.03
C(4)-S(3)	1.78 ± 0.10
C(4)-S(4)	1.66 ± 0.10
C(4)-O(2)	1.28 ± 0.14
O(2)-C(5)	1.44 ± 0.18
C(5)-C(6)	1.53 ± 0.18
Other intramolecular atom distances, Å	
Pb-S(2)	2.84 ± 0.03
S(1)-S(2)	2.98 ± 0.04
S(1)-S(4)	4.16 ± 0.04
Pb-S(4)	2.95 ± 0.03
S(3)-S(4)	2.94 ± 0.04
S(2)-S(3)	3.77 ± 0.04
Intermolecular distances	
van der Waals distance between sulfur atoms, Å	
S(2)-S(2)	3.65 ± 0.04
S(4)-S(4)	3.49 ± 0.04
van der Waals distance between methyl ends of the ethyl group, Å	
C(3') — C(3'')	4.30 ± 0.40
C(3', + b)-C(3'')	3.75 ± 0.40
C(6) — C(3'')	4.27 ± 0.25
C(6) — C(6''')	4.13 ± 0.30
C(6,+b)-C(6'')	4.47 ± 0.30
C(6,+b)-C(3'')	4.18 ± 0.25

Table 6-12 (Con't)

Other intermolecular atom distances, Å	
Pb(-b)-S(1)	3.32 ± 0.03
Pb(-b)-S(2)	4.81 ± 0.03
Pb(-b)-S(3)	3.37 ± 0.03
Pb(-b)-S(4)	3.50 ± 0.03
Bond angles, degrees	
Pb-S(1)-C(1)	87. ± 3
S(1)-C(1)-S(2)	124 ± 6
S(1)-C(1)-O(1)	113 ± 7
C(2)-C(1)-O(1)	122 ± 7
C(1)-O(1)-C(2)	118 ± 8
C(1)-C(2)-C(3)	108 ± 8
S(1)-Pb-S(3)	98.2 ± 0.9
Pb-S(3)-C(4)	91 ± 3
S(3)-C(4)-S(4)	118 ± 6
S(3)-C(4)-O(2)	116 ± 7
S(4)-C(4)-O(2)	127 ± 7
C(4)-O(2)-C(5)	117 ± 8
O(2)-C(5)-C(6)	104 ± 8

[a] After H. Hagihara and S. Yamashita, Acta Cryst,, 21, 356 (1966).

to the van der Waals distance for sulfur atoms. The distances between the Pb(-b) atom in Fig. 6-14 and the S(1), S(2), S(3), and S(4) atoms listed in Table 6-12 are larger than the sum of the divalent ionic radii of lead and sulfur atoms, 3.05 Å. This indicates that there is no direct binding between molecules piled up in the b-axis direction.

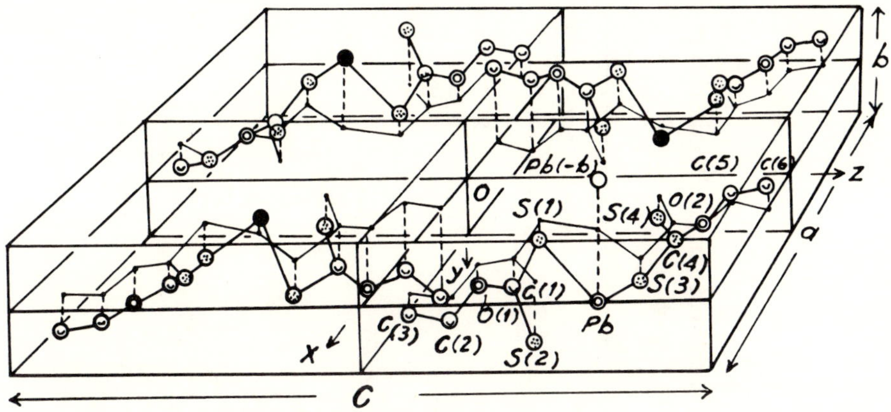

FIG. 6-14

Representation of the configuration of atoms in the unit cell of lead ethyl xanthate. Small dots denote projections of atoms on the plane y = O [After H. Hagihara and S. Yamashita: Acta Cryst., 21, 350 (1966).]

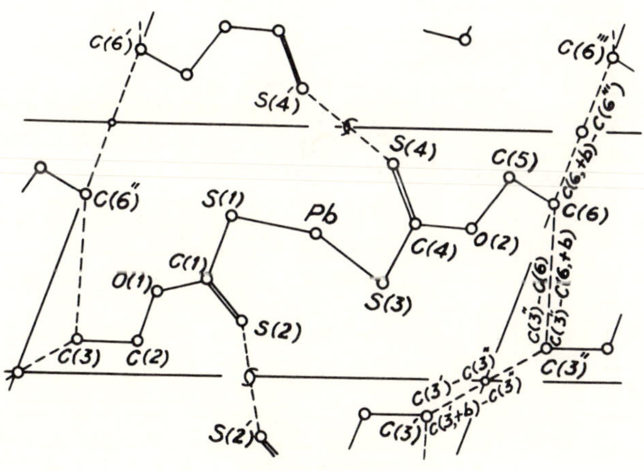

FIG. 6-15

Configuration of lead ethyl xanthate. [After H. Hagihara and S. Yamashita: Acta Cryst., 21, 350 (1966).]

TABLE 6-13

Interatomic Distances, Bond Angles, and their Standard Deviations in Lead n-Butyl Xanthate[a]

Atoms	Distance, Å	Bonds	Angle, degrees
Pb-S(1)	2.83 ± 0.04	S(1)-Pb-S(2)	65.8 ± 2.0
Pb-S(2)	2.77 ± 0.09	S(1)-Pb-S(3)	19.4 ± 1.5
Pb-S(3)	2.76 ± 0.06	S(2)-Pb-S(3)	93.7 ± 2.2
Pb-S(4)	3.03 ± 0.08	S(3)-Pb-S(4)	58.6 ± 2.0
S(1)-S(2)	3.04 ± 0.10	S(4)-Pb-S(2)	141.2 ± 2.6
S(3)-S(4)	2.85 ± 0.10	S(4)-Pb-S(1)	86.8 ± 1.7
S(2)-S(3)	4.04 ± 0.11	Pb-S(1)-C(1)	82 ± 7
S(4)-S(1)	4.02 ± 0.09	Pb-S(2)-C(1)	88 ± 7
C(1)-S(1)	1.86 ± 0.19	Pb-S(3)-C(2)	93 ± 7
C(1)-S(2)	1.62 ± 0.20	Pb-S(4)-C(2)	85 ± 7
C(2)-S(3)	1.66 ± 0.19	S(1)-C(1)-S(2)	122 ± 11
C(2)-S(4)	1.58 ± 0.20	S(3)-C(2)-S(4)	123 ± 11
C(1)-O(1)	1.25 ± 0.26	S(1)-C(1)-O(1)	122 ± 11
C(2)-O(2)	1.25 ± 0.26	S(2)-C(1)-O(1)	117 ± 16
O(2)-C(7)	1.41 ± 0.35	S(3)-C(2)-O(2)	123 ± 16
C(3)-C(4)	1.34 ± 0.42	S(4)-C(2)-O(2)	115 ± 16
C(7)-C(8)		C(1)-O(1)-C(3)	112 ± 20

Table 6-13 (Con't)

C(4)-C(5) C(8)-C(9)	1.47 ± 0.42	C(2)-O(2)-C(7)	120 ± 20
C(5)-C(6) C(9)-C(10)	1.46 ± 0.42	O(1)-C(3)-C(4)	109 ± 25
		O(2)-C(7)-C(8)	124 ± 25
		C(3)-C(4)-C(5) C(7)-C(8)-C(9)	96 ± 25
		C(4)-C(5)-C(6) C(8)-C(9)-C(10)	112 ± 25

[a] After H. Hagihara, Y. Watanabe and S. Yamashita, Acta Cryst., B24, 965 (1968).

The methyl ends of the ethyl groups of the molecules come into contact through symmetry centers as shown in Fig. 6-15. The distances between carbon atoms of the methyl group agree with the van der Waals distance for a methyl group. The deviation of the C(3) atom from the S(1)S(2)O(1) plane is interpreted by the rotation of the -C(2)-C(3) segment of the xanthate group around the O(1)-C(2) bond so that the distances between the methyl ends of the neighboring molecules become nearer to 4.0 Å while in planar configuration they are 3.4 and 4.9 Å.

A geometric interpretation is given to explain the asymmetrical molecular structure. A molecule consisting of two planar ethyl xanthate groups connected with a lead atom possesses a twofold rotation axis passing through the lead atom and normal to the plane defined by two xanthate groups. With this model the deviation of the lead atom from the plane defined by two xanthate groups is 1.80 Å, which is considerably larger than those observed in the actual molecule. When the xanthate groups are rotated by the angles shown in Fig. 6-16 (the xanthate groups rotated are distinguished by shaded circles), these distances will be 0.42 Å for one xanthate group and 0.79 Å for the second, both in agreement with the observed values. The interatomic distances Pb-S(2') and Pb-S(4'), 2.92 and 2.97 Å, respectively, are also in good agreement with the result of the analysis shown in Table 6-12.

The xanthate groups are also rotated (depicted by arrows in Fig. 6-16), each separately, by $7.5^\circ \pm 2.0^\circ$ around the Pb-S(1) and Pb-S(3) bonds. The distance between the lead atom and each xanthate group is unchanged, as are the distances Pb-S(2') and Pb-S(4'). The distances S(2)-S(3) and S(1)-S(4)

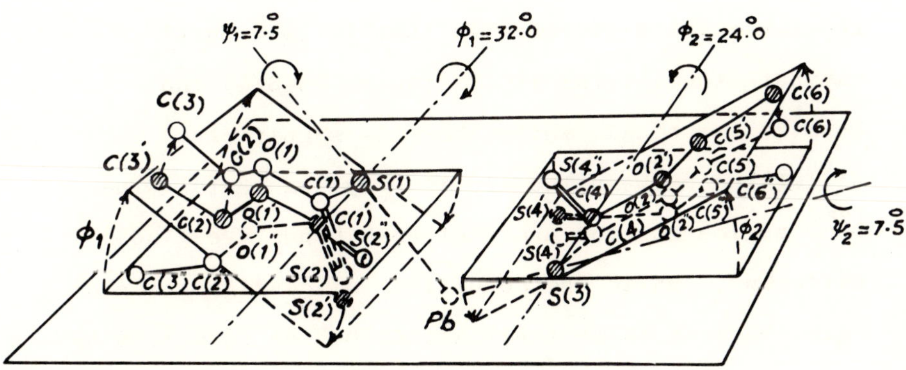

FIG. 6-16

Representation of the construction of a nonpolar asymmetric molecule of lead ethyl xanthate from a symmetrical structure by rotations of the molecular segments around C-S and Pb-S single bonds. [After H. Hagihara and S. Yamashita: Acta Cryst., 21, 350, (1966).]

change, however, the new values being 3.78 and 4.14 Å, respectively, in good agreement with the interatomic distances found by analysis. The molecular model thus constructed is shown with thick circles in Fig. 6-16. The two mutually independent rotations suggested can be explained by the van der Waals attraction between the double-bonded sulfur atoms S(4) ...S(4') and S(2) ...S(2') and the tendency for the alkyl groups of the molecule to be packed as closely as possible.

The configuration of lead n-butyl xanthate as a whole closely resembles that of lead ethyl xanthate. The important difference is that in lead butyl xanthate the interatomic distances Pb-S(1) and Pb-S(2) are equal, while there is significant difference between the two in ethyl xanthate. The bond angles for the shortest Pb-S bonds are also all different. While the conditions of configuration of lead ethyl xanthate are realized by the rotations of the molecular segments around the C-S and Pb-S bonds, as discussed above, there is no corresponding rotation in the configuration of lead butyl xanthate. The distances between sulfur atoms belonging to different xanthate groups in lead butyl xanthate are larger than the van der Waals distance between two sulfur atoms. This indicates that unlike that of lead ethyl xanthate the configuration of lead butyl xanthate is not affected by the van der Waals attraction between the sulfur atoms. The configuration of the lead atom surrounded by xanthate sulfur atoms in lead n-butyl xanthate is shown in Fig. 6-17.

Crystal structure analysis of the molecular compound lead isopropyl xanthate-pyridine, $Pb(SSCOC_3H_7)_2 \cdot C_5H_5N$ (cf. Chapter 3) has recently been described by Hagihara et al.[19] The crystal is monoclinc with a = 17.80 ± 0.03, b = 11.18 ± 0.02, c = 10.61 ± 0.02 Å; β = 107.8 ± 0.2°, and space group $P2_1/C$. The unit cell contains four formula units. The crystal is found like a sandwich consisting of alternate layers of lead isopropyl xanthate molecules and pyridine molecules piled up by van der Waals contact in the a axis direction. The lead atom is bonded to one xanthate group with Pb-S distance of 2.833 ± 0.018 and 2.896 ± 0.013 Å and to the other with 2.722 ± 0.019 and 3.057 ± 0.016 Å. The last value is the same as the sum of the ionic radii of Pb^{2+} and S^{2-}, 3.04 Å. The shortest Pb-S bond and sulfur atoms of the other

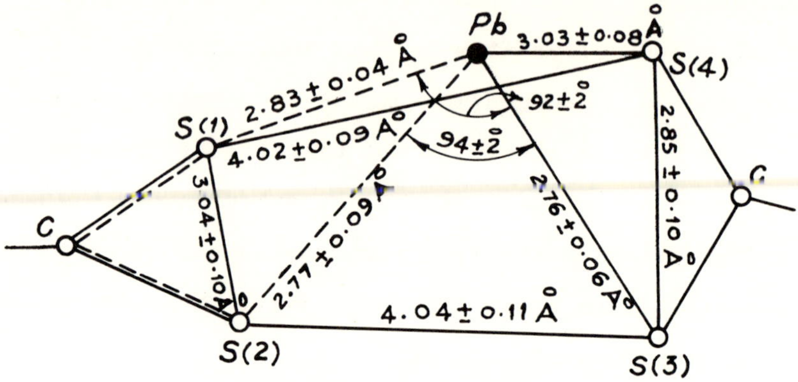

FIG. 6-17

Configuration of the four dithiocarbonic sulfur atoms about the lead atom in the lead n-butyl xanthate crystal. [After H. Hagihara, Y. Watanabe and S. Yamashita: Acta Cryst., B24, 960 (1968).]

xanthate group make S-Pb-S bonds with bond angles of $95.0 \pm 0.6^{\circ}$ and $87.5 \pm 0.5^{\circ}$.

ZINC ETHYL XANTHATE[20]

This molecule is also monoclinic, with: $a = 18.278 \pm 0.014$; $b = 5.700 \pm 0.003$; $c = 11.381 \pm 0.012$ Å; $\beta = 101.47^{\circ} \pm 0.10^{\circ}$; space groups, $P2_1/c$. Interatomic distances and bond angles are shown in Table 6-14.

Each zinc atom is surrounded by four sulfur atoms arranged almost tetrahedrally around it. Each sulfur atom belongs to a different xanthate group.

TABLE 6-14

Interatomic Distances and Bond Angles and Their Standard Deviations in Zinc Ethyl Xanthate[a]

Atoms	Distance, Å	Bonds	Angle, degrees
Zn-S(1)	2.6362 ± 0.009	S(1)-Zn-S(3)	108.7 ± 0.3
Zn-S(2')	2.337 ± 0.010	S(1)-Zn-S(4G)	110.3 ± 0.3
Zn-S(3)	2.364 ± 0.012	S(1)-Zn-S(2')	115.2 ± 0.3
Zn-S(4G)	2.369 ± 0.008	S(3)-Zn-S(4G)	107.6 ± 0.3
		S(3)-Zn-S(2')	112.1 ± 0.3
Zn-S(2)	3.420 ± 0.011	S(2)-Zn-S(4G)	102.6 ± 0.3
Zn-S(4)	3.508 ± 0.009		
		Zn-S(1)-C(1)	104.9 ± 1.0
S(1)-S(3)	3.940 ± 0.011	C(1)-S(2)-Zn"	114.8 ± 1.0
S(3)-S(4G)	3.820 ± 0.011	S(1)-C(1)-S(2)	127.7 ± 1.8
S(4G)-S(1)	3.883 ± 0.012	S(1)-C(1)-O(1)	121.9 ± 2.0
S(2')-S(4G)	3.673 ± 0.012	S(2)-C(1)-O(1)	110.3 ± 1.9
S(2')-S(1)	3.968 ± 0.013	C(1)-O(1)-C(2)	112.7 ± 2.0
S(2')-S(3)	3.899 ± 0.011	O(1)-C(1)-C(3)	102.0 ± 3.2
S(2)-S(3)	3.952 ± 0.011	Zn-S(3)-C(4)	105.5 ± 1.0
S(2)-S(4G)	3.381 ± 0.013	C(4)-S(4)-Zn(G)	105.1 ± 1.2
S(1)-S(4)	3.647 ± 0.011	S(3)-C(4)-S(4)	123.4 ± 1.6

Table 6-14 (Con't)

S(1)-S(2)	2.975 ± 0.012	S(3)-C(4)-O(2)	120.7 ± 2.3
S(1)-C(1)	1.61 ± 0.03	S(4)-C(4)-O(2)	116.0 ± 2.0
S(2)-C(1)	1.70 ± 0.03	C(4)-O(2)-C(5)	121.0 ± 2.4
C(1)-O(1)	1.42 ± 0.03	O(2)-C(5)-C(6)	106.4 ± 3.9
O(1)-C(2)	1.51 ± 0.03		
C(2)-C(3)	1.54 ± 0.05		
S(3)-S(4)	2.983 ± 0.012		
S(3)-C(4)	1.70 ± 0.03		
S(4)-C(4)	1.69 ± 0.03		
C(4)-O(2)	1.33 ± 0.03		
O(2)-C(5)	1.50 ± 0.04		
C(5)-C(6)	1.50 ± 0.06		
C(3)-C(3.S)	3.68 ± 0.06		
S(6)-C(6.1')	3.69 ± 0.07		
C(6)-C(6.1')	4.17 ± 0.08		

[a] After T. Ikeda and H. Hagihara: Acta Cryst. 21, 925 (1966).

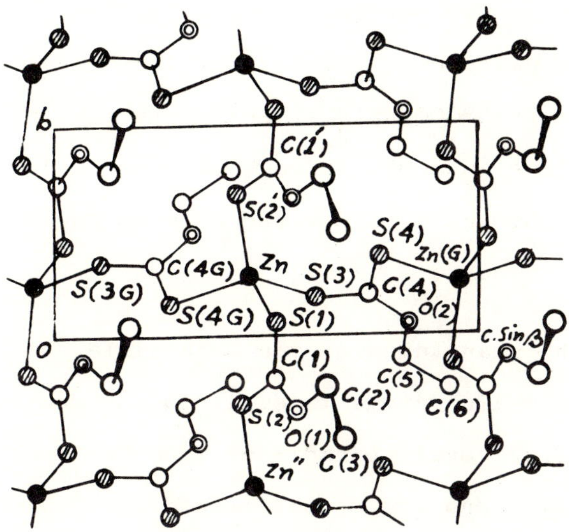

FIG. 6-18

Configuration of the atoms in the zinc ethyl xanthate crystal within the sectional layer $0 < x < 0.5$, viewed along the X-axis. [After T. Ikeda and H. Hagihara: Acta Cryst., 21, 919 (1966).]

Each xanthate group, in turn, bridges two adjacent zinc atoms in such a way that the bonding -S(1)-C'(1)-S(2)- bridges zinc atoms in the y direction and the bonding -S(3)-C'(4)-S(4)- bridges them in the z direction. The ethyl groups protrude outward on both sides of the network consisting of zinc atoms and -S-C'-S atom groups. The sheets bounded by the methyl ends of the ethyl groups are piled up in the direction through twofold screw axes. The methyl ends in successive sheets are mutually in van der Waals contact. This structure is quite different from that of lead ethyl xanthate in which $Pb(SSCOC_2H_5)_2$ is an entity that builds a crystalline assembly through van der Waals contact between the sulfur atoms and between the methyl ends in ethyl groups.

The mean value of the bond distances between zinc and sulfur atoms is equal to the sum of the tetrahedral covalent radii of zinc and sulfur atoms. Of the six bond angles, S(1)-Zn-S(3), S(3)-Zn-S(4G), and S(4G)-Zn-S(1) are almost equal to the tetrahedral angle of 109.5°. Departures from this value have been observed for the bond angles S(2')-Zn-S(1), S(2')-Zn-S(3), and S(2')-Zn-S(4G), and have been attributed to the deviation of S(2') from the apex of of the regular tetrahedron formed by the four sulfur atoms around the zinc atom.

The distances S(2')-S(4G) and S(1)-S(4) between sulfur atoms belonging to different xanthate groups are almost the same as the distance of the van der Waals contact of spherical sulfur atoms, 3.6-3.7 Å. However, the distance S(2)-S(4G) is a little shorter, possibly because of the distortion of the spherical shape of sulfur atoms as a result of their bondings with carbon and zinc atoms which permit a closer approach of S(2) and S(4G).

The results of the crystal structure analysis of zinc and lead xanthates have been applied in explaining the collector mechanism of xanthates on sphalerite (ZnS) and galena (PbS) and will be discussed in detail in Chapter 7.

CADMIUM n-BUTYL XANTHATES

The crystal structure of this compound was determined by Rietveld and Mansen[21] by three-dimensional X-ray analysis. The space group is $P2_1/a$ and the cell dimensions are: a = 11.59; b = 5.84; c = 25.7 Å; $\beta = 101^\circ 44'$. The bond lengths and bond angles are tabulated in Table 6-15.

Each cadmium atom is coordinated with four sulfur atoms in a tetrahedral arrangement with sp^3 hybridization. Some of the tetrahedral angles show large deviations from the normal value of 109°. This is believed to be caused by each sulfur atom belonging to a different xanthate radical. The Cd-S distances are in the range 2.56-2.61 Å, which is slightly longer than the expected bond distance for a pure covalent bond. These coordinate bonds form a linkage

TABLE 6-15

Bond Lengths and Bond Angles in Cadmium n-Butyl Xanthate[a]

Atoms	Distance, Å	Bonds	Angle, degrees
Cd-S(11)	2.619	S(11)-Cd-S(21)	95.3
Cd-S(12)	2.612	S(11)-Cd-S(12)	108.1
Cd-S(21)	2.560	S(11)-Cd-S(22)	118.1
Cd-S(22)	2.580	S(21)-Cd-S(22)	112.1
S(12)-C(11)	1.782	S(21)-Cd-S(12)	121.5
C(11)-S(11)	1.626	S(22)-Cd-S(12)	102.5
C(11)-O(11)	1.337	Cd-S(12)-C(11)	100.4
C(1)-C(12)	1.527	S(12)-C(11)-O(1)	113.4
C(12)-C(13)	1.489	S(11)-C(11)-O(1)	121.0
C(13)-C(14)	1.547	S(11)-C(11)-S(12)	125.2
C(14)-C(15)	1.543	C(11)-O(1)-C(12)	124.1
S(22)-C(21)	1.727	O(1)-C(12)-C(13)	99.8
C(21)-S(21)	1.640	C(12)-C(13)-C(14)	118.2
C(21)-O(2)	1.424	C(13)-C(14)-C(15)	116.7
C(22)-C(23)	1.577	C(11)-S(11)-Cd	97.9
C(24)-C(24)	1.483	Cd-S(22)-C(21)	98.9
C(24)-C(25)	1.552	S(22)-C(21)-S(21)	127.3

Table 6-15 (Con't)

C(21)-S(21)-Cd	109.3
S(22)-C(21)-O(2)	114.1
S(21)-C(21)-O(2)	118.5
C(21)-O(2)-C(22)	119.8
O(2)-C(22)-C(23)	105.1
C(22)-C(23)-C(24)	109.7
C(23)-C(24)-C(25)	107.2

[a] After H.M. Rietveld and E.N. Mansen:Acta Cryst. 18, 429 (1965).

between the xanthate groups throughout the lattice, resulting in separate layers (Fig. 6-19) which in turn are held together by van der Waals forces.

The lengths of the C S bonds differ markedly from each other, which shows that one has more

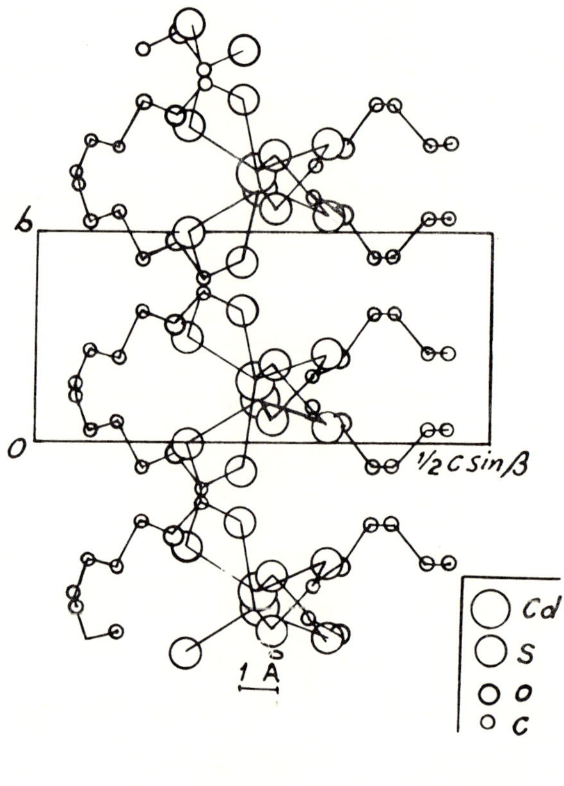

FIG. 6-19

Structure of cadmium n-butyl xanthate projected on (100), showing intermolecular coordinate linkages. [After H.M. Rietveld and E.N. Mansen: Acta Cryst., 18 429, (1965).]

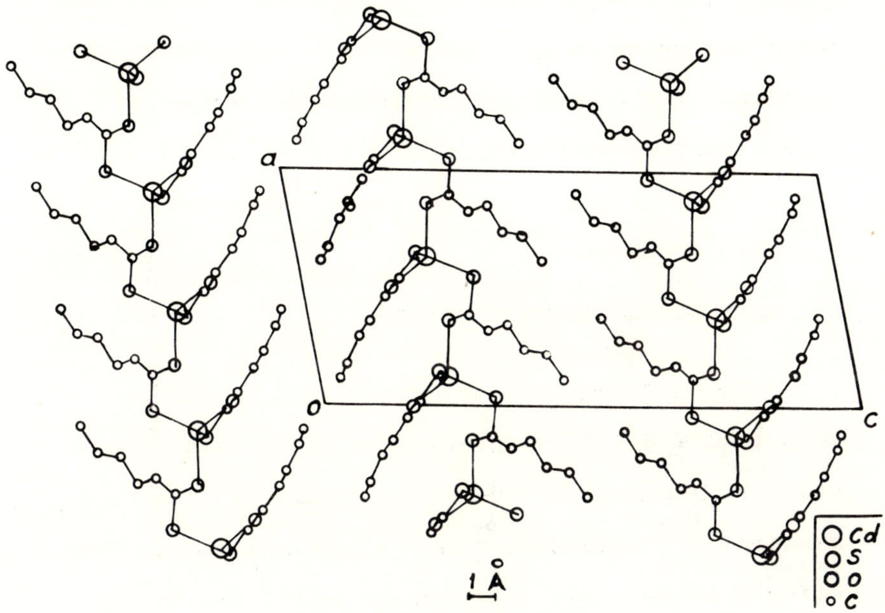

FIG. 6-20

Structure of cadmium n-butyl xanthate projected on (010), showing the intermolecular coordinate linkages. [After H.M. Rietveld and E.N. Mansen: Acta Cryst., 18, 429 (1965).]

double-bond character than the other. The average values of the C=S bond length (1.63 Å) and the C-S bond length (1.75 Å) agree with the values for the corresponding bonds found in antimonous xanthate (see Table 6-10).

The O(1)-C(12)-C(13)-C(14) chain shows a rather unusual configuration, having a cis-oriented O(1)-C(12) bond not found in xanthates of potassium, aresenic, and antimony. No explanation is available for this; there is no evidence of overcrowding when the chain would have its normal zigzag shape. It is also slightly stretched, making the C(12)-C(13)-C(14) and C(13)-C(14)-C(15) angles larger than the tetrahedral angle.

THE MOLYBDENUM-XANTHATE COMPLEX

The molybdenum complex of ethyl xanthate $Mo_2O_3(S_2COC_2H_5)$ first prepared by Malatesta (see Chapter 2) has been studied in detail by Blake and associates,[22] employing single-crystal X-ray diffraction methods. The space group is $P2_1$ and Z = 2; the unit cell dimensions are: a = 10.72 ± 0.03; b = 13.57 ± 0.03; c = 10.86 ± 0.03 Å; β = 123.5 ± 0.5°. The general structure of the molecule is shown in Figs. 6-21 and 6-22. It consists of two distorted octahedra sharing an oxygen atom so as to form a linear Mo-O-Mo group. There is a second oxygen atom on each molybdenum atom adjacent to the bridging one and the remaining four atoms bound to each molybdenum atom are xanthate sulfur atoms. The bond lengths and bond angles are listed in Table 6-16.

The crystal structure analysis is in accord with the formula originally assigned by Malatesta

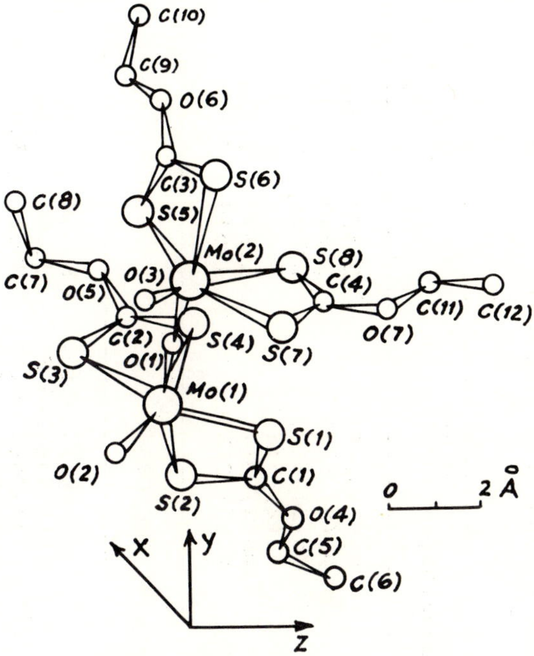

FIG. 6-21

A perspective view of the $Mo_2O_3(S_2COC_2H_5)_4$ molecule. [After A.B. Blake, F.A. Cotton, and J.S. Wood, J.Am. Chem. Soc., 86, 3024 (1964); copyright 1964 by the American Chemical Society and reproduced by permission.]

based on its diamagnetic nature ($X_m = -250 \times 10^{-6}$)

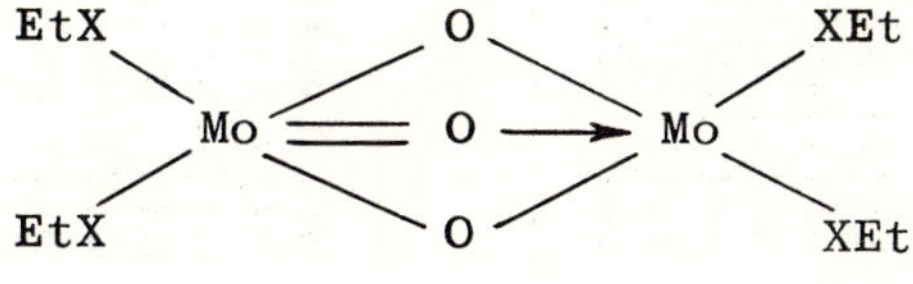

In this structure molybdenum exists both in tetravalent as well as in hexavalent states. Such

TABLE 6-16

Bond Lengths and Bond Angles around Molybdenum Atoms[a]

Bond lengths, Å			
Mo(1)-S(1)	2.458 ± 0.013	Mo(2)-S(5)	2.469 ± 0.013
Mo(1)-S(2)	2.530 ± 0.013	Mo(2)-S(6)	2.540 ± 0.013
Mo(1)-S(3)	2.509 ± 0.012	Mo(2)-S(7)	2.508 ± 0.014
Mo(1)-S(4)	2.690 ± 0.012	Mo(2)-S(8)	2.715 ± 0.012
Mo(1)-O(1)	1.851 ± 0.034	Mo(2)-O(1)	1.872 ± 0.034
Mo(1)-O(2)	1.644 ± 0.029	Mo(2)-O(3)	1.649 ± 0.028
Bond angles, degrees			
O(1)-Mo(1)-S(1)	91.48 ± 1.11	O(1)-Mo(2)-S(5)	89.56 ± 1.10
O(1)-Mo(1)-S(3)	96.11 ± 1.10	O(1)-Mo(2)-S(7)	98.89 ± 1.10
O(1)-Mo(1)-S(4)	85.71 ± 1.10	O(1)-Mo(2)-S(8)	83.11 ± 1.09
O(2)-Mo(1)-S(1)	114.84 ± 1.06	O(3)-Mo(2)-S(5)	110.23 ± 1.04
O(2)-Mo(1)-S(2)	93.20 ± 1.06	O(3)-Mo(2)-S(6)	98.85 ± 1.03
O(2)-Mo(1)-S(3)	90.04 ± 1.05	O(3)-Mo(2)-S(7)	93.95 ± 1.04
S(3)-Mo(1)-S(4)	67.27 ± 0.37	S(7)-Mo(2)-S(8)	67.40 ± 0.40
S(1)-Mo(1)-S(2)	70.46 ± 0.41	S(5)-Mo(2)-S(6)	70.91 ± 0.41

Table 6-16 (Con't)

S(1)-Mo(1)-S(4)	85.55 ± 0.39	S(5)-Mo(2)-S(8)	87.56 ± 0.40
S(2)-Mo(1)-S(4)	81.41 ± 0.38	S(6)-Mo(2)-S(8)	81.13 ± 0.39
S(2)-Mo(1)-S(3)	94.73 ± 0.40	S(6)-Mo(2)-S(7)	92.49 ± 0.43
O(2)-Mo(1)-S(4)	156.07 ± 1.05	O(3)-Mo(2)-S(8)	161.29 ± 1.04
O(1)-Mo(1)-S(2)	158.16 ± 1.11	O(1)-Mo(2)-S(6)	155.93 ± 1.11
S(1)-Mo(1)-S(3)	151.03 ± 0.40	S(5)-Mo(2)-S(7)	152.16 ± 0.45
		O(1)-Mo(1)-O(2)	105.32 ± 1.48
		O(1)-Mo(2)-O(3)	102.15 ± 1.45
		Mo(1)-O(1)-Mo(2)	178.03 ± 4.06

[a] After A.B. Blake, F.A. Cotton and J.S. Wood: J. Am. Chem. Soc., 86, 3024 (1965); (copyright 1964 by the American Chemical Society and reproduced by permission).

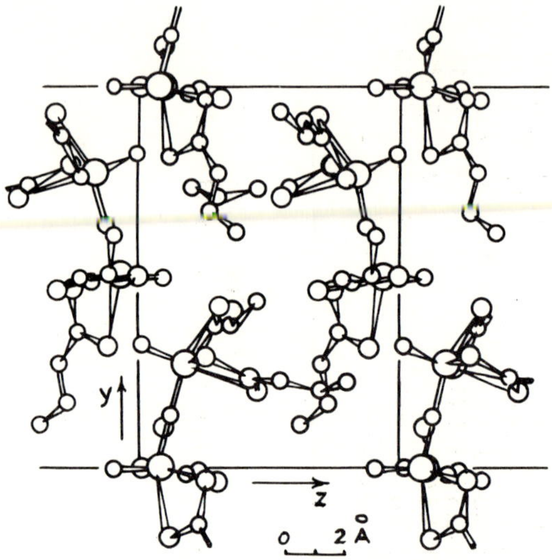

FIG. 6-22

The contents of the unit cell of $Mo_2O_3(S_2COC_2H_5)_4$ projected perpendicular to the X-axis. [After A.B. Blake, F.A. Cotton, and J.S. Wood: J. Am. Chem. Soc., 86, 3024 (1964); copyright 1964 by the American Chemical Society and reproduced by permission.]

a structure also explains the intense colors of the solutions of the complex and the metallic appearance of their crystals.

REFERENCES

1. Felumb, R.: Molecular structure of dixanthates by infrared spectroscopy. Comp. Rend., 244, 2038-2041 (1957).

2. Little, L.H. and J. Leja: Infrared spectra of xanthate compounds. Proceedings of the Second International Congress on Surface Activity, Vol. III, Butterworth, London, 1957, pp. 261-266.

3. Shankaranarayana, M.L. and C.C. Patel: Infrared spectra and the structures of xanthates and dixanthogens. Can. J. Chem., 39, 1633-1637 (1961).

4. Little, L.H., G.W. Poling, and J. Leja: Infrared spectra of xanthate compounds. Can. J. Chem, 39, 745 (1961); 39, 1783-86 (1961).

5. Rao, C.N.R., R. Venkataraghavan, and T.R. Kasturi: Contribution to the infrared spectra of organosulfur compounds. Can. J. Chem., 42, 36-41 (1964).

6. Shankaranarayana, M.L. and C.C. Patel: Infrared absorption studies on some derivatives of xanthic dithiocarbonic and trithiocarbonic acids. Spectrochim. Acta, 21, 95-103 (1965).

7. Watt, G.W. and B.J. McCormic: The infrared spectra and structure of transition metal xanthates. Spectrochim. Acta, 21, 753-762 (1965).

8. Shankaranarayana, M.L. and C.C. Patel: The electronic absorption spectra of dixanthogens. Can. J. Chem., 39, 2590-2592 (1961).

9. Shankaranarayana, M.L. and C.C. Patel: The electronic spectra of some derivatives of xanthic, dithiocarbonic and trithiocarbonic acids. Acta Chem. Scand., 19, 1113-1119 (1965).

10. Malatesta, L.: The electric moments of xanthates. Gazz. Chim. Ital., 69, 629-639 (1939); 70, 541-553 (1940).

11. Solozhenkin, P.M. and N.L. Kopitsya: E.P.R. in solutions of copper(II) xanthates. Zh. Strukt. Khim, 8(3), 534-535 (1967).

12. Mazzi, F. and C. Tadini: The crystal structure of potassium xanthate. Z. Krist., 118(5/6), 378-392 (1963).

13. Carrai, G. and G. Gottardi: The crystal structure of arsenions xanthate. Z. Krist., 113, 373-384 (1960).

14. Gottardi, G.: The crystal structure of antimonious xanthate. Z. Krist., 115, 451-459 (1961).

15. Bonatti, S.: Study of the crystal structure of nickel xanthate. Atti Soc. Toscana Sci. Nat. Pisa, Proc. Verbali M47, 71-84 (1938).

16. Franzini, M.: The crystal structure of nickelous xanthate. Z.Krist. 118(5/6), 393-403 (1963).

17. Hagihara, H. and S. Yamashtia: The crystal structure of lead ethyl xanthate. Acta Cryst., 21, 350-358 (1966).

18. Hagihara, H., Y. Watanabe, and S. Yamashita: The crystal structure of lead n-butyl xanthate. 1. Disorded structure. Act Cryst., B24, 960-966 (1966).

19. Hagihara, H., N. Yoshida, and Y. Watanabe: The crystal structure of the molecular compound lead isopropyl xanthate - pyridine. Acta Cryst. B25, 1775-1784 (1969).

20. Ikeda, T. and H. Hagihara: The crystal structure of zinc ethyl xanthate. Acta Cryst., 18, 429-436 (1965).

21. Rietveld, H.M. and G.N. Mansen: The crystal structure of cadmium n-butyl xanthate. Acta Cryst., 18, 429-436 (1965).

22. Blake, A.B., F.H. Cotton, and J.S. Wood: The crystal, molecular and electronic structures of a binuclear oxomolybdenum(V) xanthate complex. J. Am. Chem. Soc., 86, 3024-3032 (1964).

CHAPTER 7

XANTHATES IN MINERAL FLOTATION

GENERAL

Xanthates of alkali metals, sodium and potassium in particular, are very widely employed as reagents in the selective separation of sulfide minerals by the method of froth flotation. This method is particularly useful for the beneficiation of very low grade ores containing sometimes as little as 1% economic mineral. The method is based on the surface chemical reaction of certain reagents classified as mineral collectors, which create a hydrophobic film at the mineral surface. The mineral is thus rendered air avid and is carried to the air-water interface by the air bubbles produced in the pulp. Xanthates belong to the class of mineral collectors and have been in extensive use for over 40 years.

The use of xanthates as mineral collectors was introduced by Keller[1] in 1925. As the reference is available in a patent, it is not known whether the author was led to this important finding by fundamental scientific reasoning or by a purely empirical process of trial and error. In any case there seems to be no recorded work prior to 1920 on the fundamental principles that govern the selective flotation of minerals. It is now well known that a reagent that functions as a mineral collector should essentially possess a heteropolar molecular structure consisting of a polar group and a hydrocarbon group that is nonpolar. A reagent possessing this structure acts through a surface chemical reaction between the mineral and the polar group while the nonpolar group is orientated away from the mineral. The mineral surface thus acquires a hydrophobic film because of the hydrocarbon group, and this film enables the mineral particle to be carried by an air bubble.

As revealed by the chemical formula, metal xanthates possess the desired heteropolar structure comprising a hydrocarbon chain and the polar group $(S{=}C{-}S)^-$, that has a potent effect on a large number of sulfide minerals. As a result, xanthates function as efficient collectors for sulfide minerals which can thus be separated from gangue matter. Xanthates have been found to be particularly effective for galena, pyrite, and chalcopyrite, and in the presence of certain modifying agents they are used for almost the entire family of sulfide minerals. In addition, numerous attempts have been made to extend their application to flotation of some of the nonsulfide minerals, notably carbonates and sulfates.

The literature on mineral dressing[2-4] contains many references to the use of xanthates for flotation. Large quantities of copper ore occurring as sulfides of different composition such as chalcocite (Cu_2S), covellite (CuS), and chalcopyrite ($CuFeS_2$) are beneficiated by the application of xanthate collectors. In practice the xanthate is introduced in quantities that are very small relative to the quantity of the ore being treated. About 0.1 lb of xanthate per ton of ore is generally effective in obtaining the maximum possible recovery of the mineral. The reagent is introduced in the form of a dilute aqueous solution, and the ground mineral is conditioned with it by agitation by means of impellers. The pH of the pulp is adjusted to an optimum value which is 8.5-10.0 for sulfide minerals of copper.

For the flotation of iron pyrite, the pH of the pulp is generally maintained on the acid side (6.0-7.0). The difference between chalcopyrite ($CuFeS_2$) and iron pyrite (FeS_2) in optimum flotation pH is taken advantage of in the selective separation of the two minerals when they occur in the same ore body. This procedure is followed in the concentration of many deposits in which chalcopyrite is found in association with iron pyrite and is separated from the latter, as well as from associated siliceous gangue matter, by flotation by xanthates.[5]

Another important example of the use of xanthates concerns the selective beneficiation of galena (PbS) and sphalerite (ZnS). These two sulfide minerals are often found to occur in the same ore body. For the selective flotation of galena, the ore is conditioned with xanthate at a pH of about 9 and floated. Sphalerite does not float under these conditions. For the separation of the latter, the pulp remaining after the flotation of galena is treated with cupric sulfate (about 2.0 lb per ton of ore) and with an additional quantity of 0.1 lb of xanthate per ton of ore. Under these conditions sphalerite floats, leaving gangue matter consisting mostly of silica in the tailings. Cupric sulfate is said to activate the surface of sphalerite and make it responsive to xanthate.

For the collection of sulfide minerals by flotation, potassium or sodium ethyl xanthate is generally chosen as the collector. This choice is made because of the low cost resulting from the easy availability of ethyl alcohol. It is, however, recognized (as will be explained later) that the efficiency of flotation of sulfide minerals is appreciably influenced by the length of the hydrocarbon chain. The longer the hydrocarbon chain, the more efficient the collector action. In industrial practice this is not significant, however, because the recovery of the common sulfide minerals is generally as high as 95% even when only 0.05-0.2 lb of ethyl xanthate per ton of ore is employed. It is conceivable that by employing xanthates with longer hydrocarbon chains, it would be possible to reduce the quantity of collector required even further. This has in fact been shown to hold true in laboratory flotation tests,[6] but economic considerations do not favor the choice of long-chain xanthates in large-scale flotation practice except in cases in which the recovery of the mineral by ethyl xanthate is too low to be profitable.

In Table 7-1 the sulfide minerals floated by xanthate are listed along with the essential conditions for their flotation. Of the minerals indicated only the first six are of commercial interest from the standpoint of xanthate flotation at present, while the remaining ones have received attention in laboratory investigations.

It must be noted that there are no hard and fast rules that determine the choice of a particular xanthate for a particular mineral. The xanthate is generally determined by practical experience in industry and cost considerations. Based on this the choice may vary from one plant to another. The most commonly employed xanthates are listed in Table 7-1. Furthermore, the conditions of flotation are often determined by the nature of the associated minerals in the ore body. As mentioned previously,

TABLE 7-1

SULFIDE MINERALS FLOATED BY XANTHATE

Mineral	Formula	Xanthate used	Quantity of xanthate used, lb/ton of ore	Other essential conditions	Reference
Galena	PbS	Ethyl	0.05	$ZnSO_4$ (1.0 lb/ton of Ore) and NaCN (0.4 lb/ton of ore) added to prevent the flotation of the associated sphalerite	[4a]
Sphalerite	ZnS	Ethyl	0.10	$CuSO_4$ (1.0 lb/ton of ore) added as an activator	[4a]
Chalcopyrite	$CuFeS_2$	Ethyl	0.05	pH 9.0–9.5	[4b,5]
Iron pyrite	FeS_2	Ethyl	0.05	pH 6.0–6.5	[4c]

Table 7-1 (con't)

Marmatite	ZnS + FeS in varying proportion	Ethyl	0.05	$CuSO_4$ (1.0 lb/ton of ore) added as an activator	[7]
Chalcocite	CuS	Ethyl	0.05	pH 8	[4d]
Marcasite	FeS_2	a	a	–	[4e]
Realgar	AsS	a	a	–	[8]
Stibnite	Sb_2S_3	a	a		[8]
Pyrrhotite	FeS	a	a		[9]
Cinnabar	HgS	a	a		[8]
Molybdenite	MoS_2	Butyl	1.0		[10]

[a] These minerals are known to be floatable by xanthate after activation by cupric ion. The best results are generally obtained with potassium amyl xanthate; the shorter-chain xanthates are less effective. Quantities are not specified, but they are of the same order as those for the flotation of sphalerite.

by introducing certain modifying agents it is possible to make selective separation of two or more minerals by flotation. One of the conditions to be specified is pH of the flotation pulp; this is determined with a view toward accomplishing selective flotation of one mineral at a time.

There are many instances in which flotation methods have been successfully applied in the selective separation of a mineral from a complex ore body. A good example involving the use of xanthates has been described by Blaskett.[11-12] The ore under investigation was a base metal ore containing sulfides of copper iron, lead, zinc, cobalt, and nickel, in addition to a large concentration (up to 70%) of gangue matter consisting of a micaceous mineral called sericite. In a neutral or alkaline circuit employing ethyl xanthate, the recovery of sulfides was very poor. Sec-Butyl xanthate gave improved recovery, but the selectivity was poor because of the flotation of part of the sericite, which was attributed to the activation of sericite by metal ions and a consequent surface action of xanthate similar to the action of xanthate at a sphalerite surface activated by cupric ions.

This problem was overcome by conditioning the ore with xanthate in an acid circuit at a pH of 3.5, followed by flotation in a neutral or alkaline circuit. Flotation in a stronger acid circuit, although fast and clean, is not recommended because of the corrosive action of acid on flotation equipment. At pH 3.5 the concentrate recovered up to 90% of copper and included small quantities of cobalt and nickel. Blaskett[12] also attempted the selective flotation of cobalt and nickel sulfides. No satisfactory results have emerged so far.

The utility of xanthates for the collection of nonsulfide minerals has attracted considerable attention in recent years. Xanthates containing four or five carbon atoms in the hydrocarbon chain have been shown to be effective collectors for cerussite ($PbCO_3$) and anglesite ($PbSO_4$).[4f,13] They have also been tried with varying degrees of success in the flotation of less-common minerals of lead such as crocoite ($PbCrO_4$), wulfenite ($PbMoO_4$),[4f,14] and a lead vanadate mineral of indefinite composition.[15] These minerals require treatment with a larger quantity of xanthate (0.4-1.0 lb per ton of ore) than that used for the flotation of common sulfide minerals such as galena.

Longer-chain xanthates give distinctly better flotation results than ethyl xanthate. With the latter the recovery is generally incomplete.

Investigations of oxygenated minerals of copper such as malachite and azurite[48] have also shown the abstraction of xanthates by these minerals and subsequent flotation. These results, however, have not attracted attention in industry.

Another method of using xanthates in nonsulfide flotation is to treat the mineral with sodium sulfide solution prior to its conditioning with xanthate. The process, known as sulfidization creates a sulfide-like mineral surface. Treatment of such a surface with xanthate promotes efficient flotation of the mineral. This method has been investigated by Rey and co-workers[16-18] in France, and by Fleming[15] in England for the flotation of lead minerals such as cerussite and anglesite. This procedure has also been found to be of industrial value when potassium amyl xanthate is employed. Flotation of oxidized zinc ores such as smithsonite ($ZnCO_3$), hydrozincite ($ZnCO_3 \cdot Zn(OH)_2$), hemimorphite ($Zn_2SiO_3(OH)_2$, and willemite (Zn_2SiO_4) was investigated by Rey and co-workers.[18] The complex ore body also contained cerussite ($PbCO_3$). Selective separation of the different minerals required the flotation of cerussite which was achieved by sulfidizing the ore by sodium sulfide added in quantities of 0.4-1.0 lb per ton of ore and floating the sulfidized mineral by using 0.03 lb of amyl xanthate per ton of ore.

The flotation of nonsulfide minerals of copper by sulfidization and subsequent flotation by xanthate has not met with success although attempts have been made from time to time.

An interesting example of nonsulfide mineral flotation is the separation of chrysocolla, which is a complex copper silicate and, similar to silica and many other silicates, is not amenable to flotation by xanthate. However, Wright and Prosser,[19] in their study of reactions of chrysocolla with xanthates and sulfides, found that effective flotation of chrysocolla can be achieved by potassium amyl xanthate (0.2 lb/ton) following conditioning of the mineral with sodium hydrosulfide (5.0 lb/ton). Up to 99% recovery with a concentrate grade of nearly 60% expressed in terms of percentage of copper was achieved. On a practical scale, 500 g of the ore were ground in a rod mill to 90% 100-mesh B.S.S., corresponding to the complete liberation of the chrysocolla. The mineral was conditioned in a Denver 500-g flotation machine with the reagents specified above. Flotation carried out at pH 9.5 was complete in 5 min.

Flotation of native copper by xanthates was described by Fahrenwald,[20] who found that copper as coarse as 28-35 mesh floated well and minus 65-mesh copper was floated readily by ethyl xanthate. This method of flotation used in conjunction with gravity method separation is now widely adopted.[4h, 21]

Flotation of native gold by the use of ethyl and amyl xanthates in combination with other modifying flotation agents has resulted in gold recovery up to 32%.[22] The poor recovery is attributable to the presence of other associated minerals, such as iron pyrite, which are also floated by xanthate. Plaksin and Lopatin[23] reported improved recovery of gold when the alkalinity of the pulp was raised to the limit of alkali adsorption by the sulfide minerals present in the pulp.

Beneficiation of graphite by flotation using ethyl xanthate or pentasol xanthate in an medium was reported by Majumdar.[24] Although graphite possesses natural floatability, treatment with xanthate greatly enhances the recovery. The flotation of coal by xanthates has also been reported.[25] Coal dust (0-0.5 mm) is froth-floated at pH 6.5-7.0. Xanthates yield recoveries of about 97% of 7% ash content. Their effectiveness increases with the molecular weight of the alkyl radical.

FLOTATION BY LONG-CHAIN AND COMPLEX XANTHATES

In addition to the laboratory tests mentioned above, the areas of application of long-chain xanthates have been reviewed by Kakovskii et al.[26] who found that for the flotation of galena in the presence of quartz the flotation rate increases as the hydrocarbon chain length of the xanthate increases, reaching a maximum with hexyl xanthate and then decreasing. The quantity of xanthate used was 0.05 mole per ton of ore. With an increase of 0.2 mole xanthate per ton of ore, the dependence remained the same, but the difference between the higher xanthates (amyl to decyl) was less pronounced. For the flotation of sphalerite in the presence of quartz, the flotation rate increased with increasing chain length from ethyl to decyl. The results have not been explained, but they are probably correlated with the solubilities of lead and zinc xanthates.

Another set of results with complex chain xanthates has been mentioned in patent literature.[27] These xanthates have the general formula $KSCSO(CHR'CH_2O)_nR$, in which R is an alkyl, aryl, alkaryl, or cyclo alkyl group containing seven or more carbon atoms, R' is a hydrogen atom or an alkyl group containing one or two carbon atoms, and n is an integer. Laboratory tests showed that flotation time and the amount of collector needed for a mixture of galena, sphalerite, pyrite, and quartz was reduced by using a potassium alkyl xanthate obtained from an oxo-alcohol with an average molecular weight of 214 and reduced still further by using a xanthate from the same alcohol upon which 2.4 moles of ethylene oxide were fixed.

In recent years, the collector efficiency of a mixture of two or more xanthates has been investigated for sulfide minerals. For copper minerals, Anfimova and co-workers[28] found that potassium butyl xanthate mixed with potassium ethyl xanthate and potassium butyl xanthate mixed with potassium isopropyl xanthate increased the recovery. A mixture of 60% ethyl xanthate and 40% isopropyl or butyl xanthate is recommended for best results.

EFFECT OF DIXANTHOGEN ON FLOTATION

As mentioned in Chapter 3, alkali metal xanthates are prone to oxidation even in atmospheric oxygen, and the dixanthogens formed as oxidation products are also known to possess collecting properties by virtue of the presence of two alkyl groups in one dixanthogen molecule. Their collecting action on sulfide minerals had been recognized by Wark and Cox[29] and Gaudin and co-workers[30] in the course of their early studies on the collector action of xanthates, but they remained a laboratory curiosity until about 10 years ago when a study of their collector action revealed that they showed greater selectivity toward certain minerals than that shown by xanthates. The collecting properties of dixanthogens, with special reference to galena and chalcopyrite, were investigated by Arashkevich and Kakovskii.[31] Dixanthogens were emulsified before addition, and pine oil was used as a frother. For galena, the collecting ability of dibutyl dixanthogen was as good as that of butyl xanthate. Tests on a synthetic ore showed that good separation of chalcopyrite from galena can be achieved by using dibutyl dixanthogen as a collector and dichromate as a depressant for galena. This selective action of dixanthogens toward chalcopyrite has been confirmed by Rao and Patel[6], who carried out flotation studies on low-grade ore body containing chalcopyrite and iron pyrite using diethyl, diisopropyl, and diisoamyl dixanthogens introduced as an 0.03% alcoholic solution. The highest recovery (98.5%) of chalcopyrite was obtained with diisopropyl dixanthogen at a pH of 8.5. Only 0.05 lb of collector per ton of ore was used and the concentrate obtained was very much cleaner than that obtained with ethyl or isopropyl xanthate. Table 7-2 records the results obtained with different dixanthogens.

Dixanthogens could also be produced in situ in the pulp by the action of oxygen on dissolved xanthate. As discussed in Chapter 3, the oxidation of xanthate by oxygen is considerably enhanced when carbon dioxide is introduced with oxygen. Rao and Patel[6] also conducted flotation tests in the presence of different gaseous systems that have oxidizing action on xanthates. It was possible

TABLE 7-2

Flotation of Chalcopyrite by Dixanthogens[a]

Dixanthogen used	Initial pH of the pulp	Analysis of the float, %			Chalcopyrite floated, %	Iron pyrite floated, %
		Copper	Iron	Sulfur		
Diethyl	8.5	24.5	27.7	29.3	88.3	16.4
Diisopropyl	8.5	30.1	29.4	32.3	98.5	6.1
Diisoamyl	8.5	31.2	30.3	33.6	80.7	4.5
Diethyl	6.5	19.5	24.9	25.8	86.9	27.3
Diisopropyl	6.5	23.9	27.8	30.0	91.2	22.5
Diisoamyl	6.5	30.2	30.1	32.2	80.7	4.5

[a] Collector used: 0.005 lb per ton of ore. Rate of aeration of the flotation pulp: 0.25 liter/min. Time of aeration: 15 min. Frother: 0.015 lb pine oil per ton of ore.

to obtain over 90% recovery using only 0.005 lb of xanthate per ton of ore, which is nearly one-eight of the quantity of xanthate currently used in industrial practice.[5] Table 7-3 records the results obtained by Rao and Patel.[6]

A comparison of the two sets of results (Tables 7-2 and 7-3) reveals that concentrates obtained using pure dixanthogens yield a better grade of chalcopyrite.

In industry dixanthogens are given the trade name "Minerec." Their potentialities for the selective separation of sulfide minerals have not been fully exploited although their important role in collector action has long been recognized. In fact, as will be discussed in the next two sections, the presence of dixanthogen in conjunction with xanthate is considered essential for the collector action on sulfide minerals. This is at present of great theoretical significance.

CONTACT ANGLE STUDIES ON MINERALS TREATED WITH XANTHATES

In their classic investigations on flotation fundamentals, Wark and Cox[29] determined the contact angle made by an air bubble at a mineral surface

TABLE 7-3

Effect of Different Gaseous Systems on the Flotation of Chalcopyrite by Xanthates at pH 8.5

Xanthate used	Gaseous system employed	Analysis of the float, %			Chalcopyrite floated, %	Iron pyrite floated, %
		Copper	Iron	Sulfur		
Ethyl xanthate	Air	24.3	27.2	31.3	97.2	26.9
	Oxygen	25.6	29.2	33.6	96.1	25.4
	Air + CO_2	27.1	30.8	33.9	77.8	18.7
	Oxygen + CO_2	25.9	30.7	33.3	83.0	24.5

Isopropyl xanthate	Air	22.5	27.3	30.2	99.5	33.2
	Oxygen	22.8	27.4	30.5	99.9	33.8
	Air + CO_2	25.0	29.2	30.7	99.8	21.4
	Oxygen+CO_2	24.8	28.9	30.9	99.8	23.8
Isoamyl xanthate	Air	20.8	26.7	28.6	99.8	36.3
	Oxygen	20.6	26.7	28.9	100	39.1
	Air + CO_2	18.8	25.8	24.7	100	30.6
	Oxygen + CO_2	18.7	25.9	25.3	100	34.4

conditioned with xanthate solution (25 mg per liter). It is known from thermodynamics[4i] that the magnitude of the contact angle is a measure of the displacement of a solid-water interface by a solid-air interface and, hence the greater the magnitude of contact angle, the higher the flotation recovery expected. Wark and Cox[29] determined the increase in contact angle with increase in the length of hydrocarbon chain; their results, plotted in Fig. 7-1, show the values of characteristic equilibrium contact angles for different hydrocarbon chains. The values of contact angles observed lead to the inference that the efficiency of the collector should increase with the molecular weight of the alkyl group. Again, the branched chain xanthate that shows a higher contact angle than a straight-chain xanthate of the same molecular weight would be expected to show better collector efficiency; as an example, a mineral conditioned with isopropyl xanthate shows a contact angle of 74°, while that with n-propyl xanthate shows a value of 67°. Both these facts are borne out in flotation practice.

Wark and Cox[29] have also investigated the effect of pH on air bubble contact angles. They

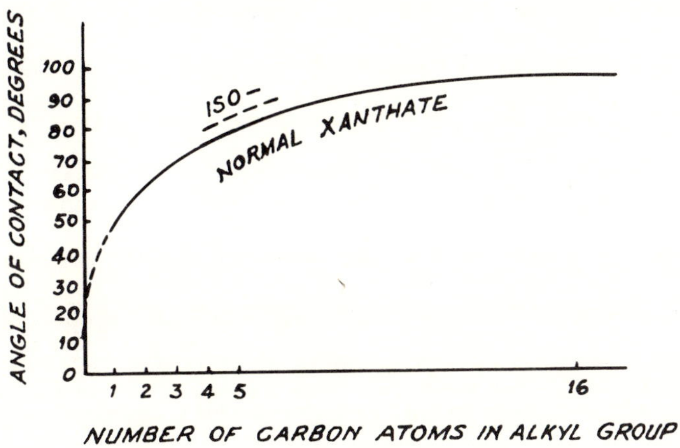

FIG. 7-1

Dependence of contact angle on the hydrocarbon chain of alkyl xanthate. [After I.W. Wark and A.B. Cox: Trans. AIME, 112, 189 (1934).]

have established the existence of a critical pH for each mineral the value of which depends upon the concentration of xanthate and the length of hydrocarbon chain. Above this pH the mineral fails to respond to xanthate because of the high concentration of hydroxyl ions which prevent the adsorption of xanthate at the mineral surface. At the points corresponding to the critical pH for various concentrations of xanthate, the product $[H^+][X^-]$ = constant. This is known as the Barsky relationship, which can be expressed in the form

$$\frac{[X^-]}{[OH^-]} = \text{a constant}$$

This implies competition between xanthate anions and hydroxyl anions for the mineral surface. If the concentration of hydroxyl ions exceeds a certain value for a particular concentration of xanthate, xanthate ions are prevented from being adsorbed at the mineral surface. This factor is utilized in the selective flotation of sulfide minerals through pH control, as mentioned before (See Table 7-4).

The limitation of the concept of critical pH as a fundamental factor in flotation has been noted by Fleming[32] in the course of his studies on the effects of alkalinity on the flotation of lead minerals. He found that when the collection of cerussite ($PbCO_3$) is carried out in solutions containing an appreciable amount of soluble carbonate the alkali exerts its depressant effect indirectly through control of the carbonate ion concentration. Another cause of depression of galena by alkali could be through the decomposition of the xanthate coating by alkaline solutions.

TABLE 7-4

Critical pH Values for Sulfide Minerals in Potassium Ethyl Xanthate Solution (25 mg/liter)[a]

Mineral	Chemical composition	Critical pH
Sphalerite	ZnS	Does not respond at any pH value
Pyrrhotite	FeS_2	6.0
Arsenopyrite	$FeAsS_2$	8.4
Galena	PbS	10.4
Pyrite	FeS_2	10.5
Marcasite	FeS_2	11.0
Chalcopyrite	$CuFeS_2$	11.8
Covellite	CuS	13.2
Bornite	Cu_5FeS_4	13.8

Table 7-4 (Con't)

Chalcocite	Cu_2S	Above 14.0

[a] From I.W. Wark and A.B. Cox: AIME Tech. Publ. No.461 (1932); Trans. AIME, 112, 189-244 (1934).

Fleming suggests that the conclusions he arrived at for the xanthate-galena system may not hold true for other sulfide minerals in which the depression may follow an entirely different mechanism. When alkali depression is caused by the decomposition of collector coatings, the term critical pH can only be relative; it varies with the time allowed for the decomposition. Furthermore, if the sulfide mineral surface is superficially oxidized, thicker coatings of xanthate are formed as a result of increased metal ion concentration at the mineral surface, and this leads to greater tolerance of the depressant action of cyanide and alkalis .

THEORIES OF THE COLLECTOR ACTION OF XANTHATES

In the early years of investigations on the physicochemical aspects of flotation, there was a considerable amount of confusion regarding the collector mechanism of xanthates. A large number of theories, many of which were based on insufficient experimental evidence, were put forward to explain the mechanism of the collector action of xanthates. As more precise methods of investigations were developed in the field of flotation, most of these theories were found to be in disagreement with the experimental observations and were gradually discarded. The only two theories that were retained to explain the action of xanthates in a number of sulfide minerals, and which promoted intensive fundamental research in the 1930's, were the solubility theory and the ion adsorption theory.

SOLUBILITY THEORY

This theory, developed by Taggart,[2,33] postulates that the collector film on the mineral surface is formed by chemical reactions of well-known types. Furthermore, the compound formed at the mineral surface is assumed to be less soluble than the same compound in the bulk. The validity of this theory requires that the sulfide mineral be superficially oxidized, giving rise to oxysulfur compounds of the metal, and that these in turn double decompose with alkali metal xanthate, forming a film of the heavy metal xanthate which renders the mineral hydrophobic. As will be discussed in the next section, the galena-xanthate system appears to satisfy these requirements. According to Taggart's theory, the nonfloatability of sphalerite by ethyl xanthate is attributed to the comparatively high solubility of zinc ethyl xanthate, which prevents the formation of a hydrophobic film on sphalerite. If, however, sphalerite is activated by cupric sulfate, it is rendered hydrophobic as a result of the formation of cupric or cuprous xanthate which has a low solubility.

However, many of the facts observed in flotation practice and also in fundamental investigations are not in accord with the solubility theory. Wark and Cox[34] have pointed out that if the solubility theory were valid pyrite and marcasite, both of which have a similar chemical composition (that is FeS_2, but different crystalline structure), could be expected to form the same kind of chemical compound with xanthate and should float under identical conditions; but in practice this is not true. Flotation of native gold by xanthate, mentioned in the previous section, cannot be explained by the solubility theory since a chemical compound of gold with a xanthate group is not expected, considering the principles of chemical bonding. A similar situation occurs in the case of graphite floated by xanthate. Taggart's theory cannot explain the selective flotation of minerals by xanthate as noted by Cook and Nixon.[35]

ION ADSORPTION THEORY

The formation of a hydrophobic film on a mineral surface conditioned with xanthate has been explained by Wark and Cox,[36,37] and later by Gaudin and co-workers,[38] to be attributable to the exchange adsorption of xanthate ion on the mineral surface. The exchange may be between the lattice ions or gegen ions of the electrical double layer on the mineral surface and xanthate ions in the solution. Wark's theory assumes that as soon as a new mineral surface is created by crushing or grinding of the mineral, it becomes a seat of adsorption of ions from water or from the surrounding medium. When xanthate is added, xanthate ions are adsorbed in place of the ions already present on the surface of the mineral. This theory of exchange adsorption accounts for the orientation of the xanthate film, one part (polar group) of which is attached to the mineral surface and the other part of which points away from it. Such orientation cannot be expected in normal chemical compounds.

As will be discussed in the next section, the ion adsorption theory is also in accord with a number of experimental findings with various minerals floated by xanthates. In fact the ion-exchange mechanism is today widely accepted as an explanation of the collector action of xanthates.

NEUTRAL MOLECULE ADSORPTION THEORY

In 1950 Cook and Nixon[35] proposed a somewhat different mechanism for the formation of a hydrophobic film on the mineral surface. According to this theory, the alkali metal xanthate dissolved in the solution undergoes hydrolysis, giving rise to xanthic acid which is adsorbed at the mineral surface. The authors of this theory argue that if only collector ions are adsorbed, without the counterions, a large amount of charge will accumulate on the mineral surface when a monolayer of collector ions is formed. As this would be an abnormal situation, the only possible way in which ionic adsorption can occur is by the simultaneous adsorption of counterions along with xanthate ions, that is, potassium or sodium ions. Adsorption of ion pairs is questionable, however, because this would give rise to a hydrophilic surface resulting from the outward orientation of potassium ions while xanthate ions are oriented toward the mineral. The basic feature of this theory is its disagreement with the ion-exchange theory. That neutral molecule (xanthic acid) adsorption at a galena surface is in accord with fundamental free energy considerations has been recently substantiated by Cook.[39]

The theory of adsorption of xanthic acid was supported by Hagihara,[40] who investigated the nature of the xanthate films on galena by the electron diffraction method. Hagihara has inferred that when the concentration of xanthate is about 500 mg per liter Y films are formed with lead atom as the center of adsorption. He further concludes that if the xanthate ion were to be adsorbed, the compensating adsorption of cations, namely, sodium or potassium ions, would be difficult to account for because of spatial considerations, that is, there is no room to accommodate them in the neighborhood of the xanthate ions. Hence, either hydrogen ions, which are very much smaller than the alkali metal ions, are adsorbed to counteract the charge of the xanthate ions, or neutral xanthic acid molecules are adsorbed on the galena lattice. The instability of the films on exposure of the mineral specimen has led Hagihara to believe that the probable mechanism involves the adsorption of xanthic acid. Hagihara's conclusion is also based on a fundamental disagreement with the ion-exchange mechanism.

The conclusions of Hagihara have been criticized by Bhrany and Rao,[41] who have pointed out that the spatial difficulties encountered by Hagihara cannot be removed, even if it is assumed that the xanthate ion alone is adsorbed, as the cross-sectional area of butyl xanthate is larger than that of a lead and sulfur pair on the galena lattice. The spatial considerations only indicate that each molecule of xanthate may occupy more than one lead and sulfur pair at the surface of galena. The instability of the films observed by Hagihara does not necessarily imply the adsorption of xanthic acid, as the alkali xanthates are also well known for their instability in air and water (cf. Chapter 3).

Thus, the neutral molecule adsorption theory has not been firmly established even in the case of galena and xanthate collectors. As will be discussed in the next section, most of the fundamental studies on adsorption and surface reactions of xanthate at galena surfaces support the ion-exchange mechanism rather than the neutral molecule adsorption mechanism. However, fresh evidence has been very recently advanced on the collector mechanism in sphalerite which supports Cook and Nixon's theory. A detailed discussion will be given later.

PLAKSIN'S VIEWS ON THE MECHANISM OF THE COLLECTOR ACTION OF XANTHATES

Plaksin and his co-workers in the USSR have carried out a series of investigations on the effect of dissolved gases on the floatability of sulfide minerals by xanthates[42-45] and have drawn conclusions about the collector mechanism of xanthates that are of great theoretical significance. The results of their investigations, some of which will be referred to in the next section, have broadly established that oxygen plays an essential role in the adsorption of xanthate. This aspect will be discussed with special reference to galena, but a similar mechanism is believed to hold true for all the polar sulfide minerals although each ore has not been investigated as extensively as galena.

ADSORPTION AND SURFACE REACTIONS OF XANTHATES ON DIFFERENT MINERALS AND THE COLLECTOR MECHANISM

GALENA

More than any other mineral, galena (PbS) has been the choice of many investigators for fundamental studies on the action of xanthate collectors. The wide variety of physiocochemical investigations carried out on this mineral during the last 4 decades have culminated in theories on the mechanism of the collector action of xanthates (surveyed in the preceding section). The validity of these theories for several other minerals has also been examined and although it is not possible to draw very general conclusions, reasonable explanations have been given for the collector mechanism of xanthates for different minerals tested.

As mentioned before, one of the oldest theories concerning the collector action of xanthates postulates that the mineral is superficially oxidized and the resultant sulfoxylate ions at the surface displaced by xanthate ions to form a lead xanthate film. This reaction was first studied by Taylor and Knoll[46] who observed that when galena was in contact with a solution of alkali metal xanthate the xanthate radical displaced sulfoxylate ions, the order of displacement of ions being roughly parallel to the respective solubilities of the salts in water. A satisfactory stoichiometric balance was found. Their data, recorded in Table 7-5, indicate the surface reaction:

$$\underset{\boxed{\text{Galena}}}{Pb^{2+}\ 2A^-} + \underbrace{2K^+ + 2X^-}_{\text{In solution}} \longrightarrow \underset{\boxed{\text{Galena}}}{Pb^{2+}\ {+2X^-}} \ \underbrace{+2K^+ + 2A^-}_{\text{In solution}} \qquad (7\text{-}1)$$

in which A^- denotes the anions initially present on the galena surface and X^- denotes the xanthate ions. The total quantity of A^- participating in the exchange reaction slowly increases as the concentration of xanthate ion in the solution is increased. Furthermore, the greater the extent to which the mineral is oxidized, the higher the quantity of A^-. This fact has been confirmed by Herd and Ure,[47] who found a decrease in the exchange capacity of galena in the absence of oxygen and concluded that when oxygen is rigidly excluded clean galena shows no exchange ability and is not

TABLE 7-5

Stoichiometry of Surface Reaction of Xanthate at Galena[a,b,c]

Xanthate ion abstracted	Reducing sulfoxylate ions (e.g. thio-sulfate)	Ions emitted[c]		
		Sulfate	Carbonate	Total
39.0	8.0	5.8	26.0	39.8
58.3	12.8	16.6	27.2	57.1
64.8	14.1	18.7	25.8	58.6
88.8	19.6	27.1	41.0	87.7

[a] After T.C. Taylor and A.F. Knoll, Trans. AIME, 112, 382-397 (1934).

[b] Initial concentration of potassium ethyl xanthate was 200 mg.

[c] All ions are expressed as milligrams of potassium ethyl xanthate per liter or the stiochiometric equivalent.

floated by xanthates. This conclusion has been supported by Knoll and Baker.[48] The results of these early investigations were explained by Taggart and Hassialis,[49] who inferred that the formation of a lead xanthate coating on the surface of galena followed the law of mass action and that the xanthate coating occurred when the solubility product of the coating compound was exceeded.

Gaudin and Wilkinson[50] observed that when galena reacted with an alkali metal xanthate lead xanthate was formed which readily decomposed to sulfur and an unidentified oil. In a later study Gaudin and his collaborators[30] found that an oxidized surface of galena reacted with an alkali metal xanthate to give a coating of lead xanthate which on prolonged grinding yielded sulfur and unidentified oils, presumably dixanthogens formed by the oxidation of xanthate.

The surface reactions of xanthate with galena have been recently reconsidered by Mellgren[51] who determined the enthalpies of adsorption of potassium ethyl xanthate at a galena surface and compared them with the heats of reactions between ethyl xanthate and lead salts. The calorimetric measurements showed that ethyl xanthate reacts with lead carbonate, basic lead carbonate, lead thiosulfate, and lead sulfate according to the principle of double decomposition, but when ethyl xanthate was made to react with lead hydroxide at a high pH, basic lead thiosulfate, lead sulfite, and lead metasulfite there were secondary reactions which gave rise to abnormal values, for the heat of reaction. The enthalpies of adsorption of xanthates at a galena surface were determined for a series of galena samples conditioned with different sulfoxylate ions such as sulfate and thiosulfate. Some of the values obtained by Mellgren[51] are shown in Table 7-6 together with the heats of reactions between potassium ethyl xanthate and the corresponding lead salts.

The close agreement between enthalpies of adsorption of xanthate at a galena surface conditioned with sulfoxylate ions and carbonate ions and heats of reactions between xanthate and the corresponding lead salts is in accord with the results of Taylor and Knoll[46] mentioned above. Examined superficially, these findings tend to support the chemical compound theory of a collector mechanism, according to which the formation of a hydrophobic film is the result of chemical reactions of well-known types. There are some hindrances to the drawing of this conclusion, however. The

TABLE 7-6

Enthalpies of Adsorption of Potassium Ethyl Xanthate at Galena Conditioned with Sulfoxylate Ions and Heats of Reactions between Ethyl Xanthate and Lead Salts[a]

Reaction	$-\Delta H$, kcal/mole Pb^{2+}
1. $[\text{Galena}]SO_4 + 2KX$	
$[\text{Galena}]_{X}^{X} + K_2SO_4$	
$PbSO_4 + 2KX \longrightarrow PbX_2 + K_2SO_4$	22.6
2. $[\text{Galena}]_{SO_4}^{SO_4} + 4\ KX \longrightarrow$	
$[\text{Galena}]_{X\,X}^{X\,X} + 2K_2SO_4$	14.8
(Double sulfate formed by treatment with 0.01 M K_2SO_4 Solution)	
$\frac{1}{2}.2PbSO_4 \cdot K_2SO_4 + 2KX \longrightarrow PbX_2 + 3/2K_2SO_4$	
$PbX_2 + 3/2K_2SO_4$	14.6
3. $[\text{Galena}]_{S_2O_3}^{OH} + 3KX \longrightarrow$	

Table 7-6 (Con't)

[Galena] X X X + $K_2S_2O_3$ +KOH	19.7
(Basic thiosulfate formed by treatment with 0.001 M $K_2S_2O_3$ at pH 8)	
$\frac{1}{2}PbS_2O_3 \cdot Pb(OH)_2 + \frac{1}{2}K_2S_2O_3 + 2KX \longrightarrow$	
$PbX_2 + \frac{1}{2}K_2S_2O_3$	19.6
4. [Galena] CO_3 + 2KX $\longrightarrow$	
[Galena] X X + K_2CO_3	17.1
$PbCO_3$ + 2KX $\longrightarrow$	
$PbX_2 + K_2CO_3$	17.2

[a] After O. Mellgren, Trans. AIME, 235, 46-50 (1966).

first is that the galena samples employed by Mellgren were treated with sulfoxylate ions such as sulfate and thiosulfate and were not deliberately oxidized. This means that the ions were present at the surface and that the adjoining electrical double layer with the counterions, namely, potassium ions was anchored in the solution.

Second, if the formation of normal lead xanthate were the only cause of the creation of a hydrophobic film, good floatability would be expected for lead xanthate precipitated in a beaker. Lead ethyl xanthate precipitate in a qualitative test showed poor floatability, however. Furthermore, Mellgren has shown that the enthalpy of adsorption of xanthate, even when a very low concentration of xanthate is employed, corresponds to the heat of reaction between xanthate and lead salt. At such low concentrations the quantity of lead xanthate formed is so small that the normal solubility of lead ethyl xanthate would have precluded its formation. These considerations indicate that the adsorbed species present at the mineral surface is not identical with the lead xanthate in bulk. In fact this difference between the two had been recognized by the authors of the chemical compound theory, who suggested that lead ethyl xanthate in bulk formed at the surface has a much lower solubility than lead xanthate in bulk. The thermochemical data of Mellgren provide thermodynamic reasoning in support of this postulation. It is inferred that although the binding energy of the anion attached to the galena surface is the same as that between an anion and a lead ion in a corresponding lead salt, the entropy content and the Gibbs free energy of the adsorbed species are different from those of the lead salt in bulk. The formation of a hydrophobic film as such is the result of an ion-exchange mechanism between anions present at the surface and in the double layer and xanthate ions present in the solution. This implies that the lead ions present form a bridge between galena and the anions, and galena thus functions as an ion-exchange medium.

ADSORPTION STUDIES ON GALENA

The mechanism of the collector action of xanthate at a galena surface has also been investigated in a number of adsorption studies. Bushell[7] has reported a detailed investigation on galena from the Sullivan mine in Canada. It was shown that the adsorption of xanthate on galena roughly follows a linear relation between log $m^2[H^+]$ and log S, where m is the equilibrium concentration of xanthate and S is the xanthate uptake. Furthermore, continuous uptake of xanthate occurred, indicating that the mineral acquires a multilayer film of xanthate. It was inferred by Bushell[7,52] that adsorption in the first layer is different from adsorption in subsequent layers. In the first adsorbed layer, the polar parts of the xanthate ions are attached to the mineral with their hydrocarbon ends oriented toward the solution, thus rendering the surface hydrophobic. In subsequent layers xanthate is attached to xanthate, the bonding being different from that between xanthate and the mineral. The xanthate in the built-up layers is oriented with hydrocarbon ends always facing the solution, which explains the fact that no decrease in hydrophobicity is observed as the equilibrium concentration of the xanthate increases or the pH decreases. The linear increase in the xanthate uptake with $m^2[H^+]$ can be explained by suggesting that the unit structure in a multilayer consists of two xanthate ions, one hydrogen ion, and one sodium or potassium ion to maintain the electrical neutrality. The multilayer of adsorbed xanthate is schematically represented in Fig. 7-2.

Burkin and associates[53] found that ethyl xanthate and undecyl xanthate ions in a strongly reducing solution adsorb reversibly at a galena surface and that the xanthate uptake is increased if the mineral is treated with lead ions and

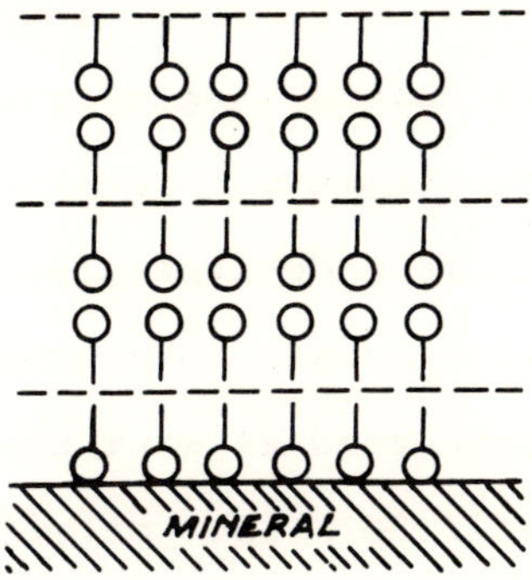

FIG. 7-2

Schematic drawing of adsobed xanthate. [After C.H.G. Bushell: Can. Mining Met. Bull., 61, 65-77 (1958).] The straight-line "tails" represent hydrocarbon chains.

subsequently washed. The results are in agreement with thermodynamic considerations since it was shown by the same authors that no reaction between lead sulfide and xanthate can occur unless the surface of lead sulfide contains a stoichiometric excess of lead. In nonreducing solutions chemical reaction occurs between xanthate and galena in accordance with the findings of Taylor and Knoll.[46]

The adsorption of potassium ethyl xanthate by galena using a column technique was recently reexamined by Mellgren and Subba Rau.[54] They studied adsorption in a system substantially free of oxygen and also determined the effect of oxidation of the galena sample. A galena sample oxidized with hydrogen peroxide showed a large increase in xanthate uptake. In a deoxygenated system the maximum uptake of xanthate corresponded to nearly a monolayer and occurred at xanthate concentrations higher than about 1.56×10^{-4} mole/liter. The data did not fit a Langmuir adsorption isotherm and the authors concluded that there is no simple relationship between adsorption density, Γ xanthate concentration, and pH for the galena-xanthate system although the data could be expressed by the equation

$$\log \Gamma = \log \Gamma^{o} + S(pH - pH^{o}) \qquad (7\text{-}2)$$

where $\log \Gamma^{o}$ and pH^{o} are constants independent of concentration and S is a function of concentration.

The rate of adsorption of xanthate in a stationary bed of galena was expressed by the equation

$$\frac{d\Gamma}{dt} = k\left[1 - \frac{\Gamma}{\Gamma^{o}}\right]^{2} cf \qquad (7\text{-}3)$$

where Γ = adsorption density at time t, Γ^o = saturation adsorption density for a given concentration, c = concentration of influent, f = flow rate, and k = a constant dependent on units. This equation was found to hold true when Γ / Γ^o was less than 0.7.

The results obtained with a stirred bed for surface coverage less than 60% followed the equation

$$\frac{d\Gamma}{dt} = k\left[1 - \frac{\Gamma}{\Gamma^o}\right] cf \qquad (7\text{-}4)$$

The authors[54] recognized that the rate of adsorption is controlled by the rate of diffusion of xanthate at low surface coverages, but at high surface coverages surface reaction controlled the adsorption rate.

The results obtained by Mellgren and Subba Rao[54] indicating a monolayer adsorption of xanthate appear to contradict the multilayer adsorption observed by Bushell.[7,52] This discrepancy was reconsidered by Gaudin and Finkelstein,[55] who pointed out that it was attributable to the considerably higher flow rates (100-200 ml/min) used in

Mellgren's work. This difference in flow rates made a 200-fold difference in the residence time and, hence, the sensitivity of detection of the reaction was very much reduced. When low flow rates (0.3 - 1.5 ml/min) were employed by Gaudin and Finkelstein, however, they found that the effluent concentration remained significantly different from that of the head solution even after the flow of xanthate solution was continued for several days, thus confirming the multilayer adsorption reported by Bushell. These findings also corroborated the conclusion drawn by earlier workers[47,48] that no interaction takes place between galena and xanthate in the absence of oxygen.

Gaudin and Finkelstein[55] further investigated the effect of controlled additions of oxygen on xanthate uptake by galena. Their results, illus- illustrated in Fig. 7-3, show that after an initial induction period during which there was no strong interaction (indicated by the steady rise in the slopes of the plots) the system eventually settled down to a continuous abstraction of xanthate from the solution at a finite rate which decreased slowly with time but never became constant within

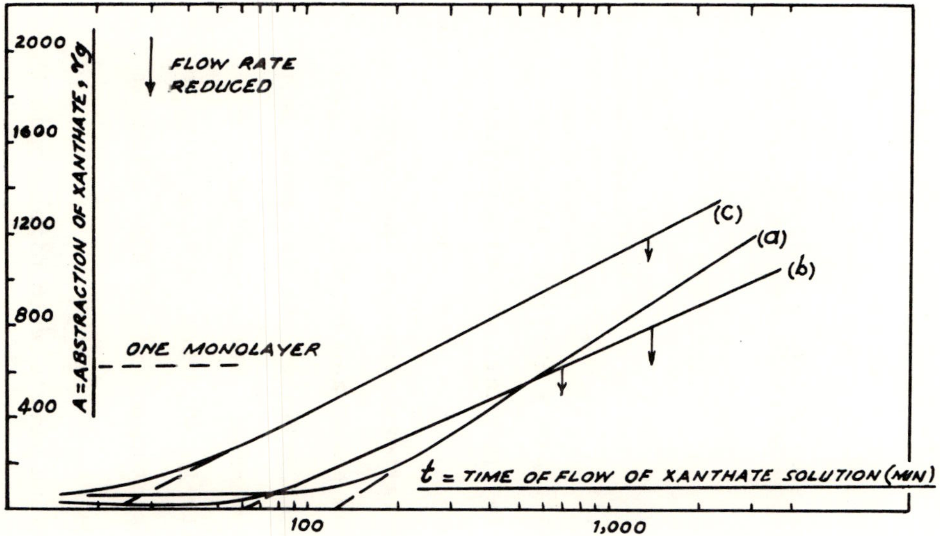

FIG. 7-3

Relationship between the quantity of xanthate abstracted from solution (A) and the time of contact with galena (t).(a) 0.6 ppm Oxygen in xanthate solution only, (b) 0.3 ppm oxygen in both rinse and xanthate solutions; (c) 0.6 ppm oxygen in both rinse and xanthate solutions. [After A.M. Gaudin and N.P. Finkelstein: Nature, 207, 389-91 (1965).]

the period of experiment. The system obeyed the equation

$$A = D \ln t + C \tag{7-5}$$

where C and D are constants and A denotes the quantity of xanthate abstracted from the solution in time t.

It has also been noted that the mineral-xanthate interaction is greatly enhanced by contact with oxygen. This effect of oxygen had been reported by several investigators earlier, and there has been considerable speculation that multilayers of adsorbed film consist of a mixed film of xanthate and dixanthate, the latter formed by the oxidation of xanthate by oxygen catalyzed by the metal ions present in the mineral lattice. As far back as in 1934, Gaudin and co-workers[30] recognized that the oxidation of xanthate in solution, giving rise to neutral molecule dixanthogen, might play a crucial role in the creation of the hydrophobic film. Their observation that galena reacts with dixanthogen to yield lead ethyl xanthate was ascribed to the reduction of dixanthogen by lead sulfide a fact later confirmed by infrared studies; see pp. 290–293. The reaction at a galena surface was found to be accompanied by the formation of a benzene-soluble red-brown precipitate, which decomposed into lead xanthate, sulfur, and lead sulfide. These authors suggested that dixanthogen formed by the oxidation of xanthate acts as the effective adsorbate species. Again in 1952 Schulman[56] expressed this opinion at the International Mineral Dressing Congress, London, and since then much experimental evidence has accumulated in support of this view although there is still no unanimity on the exact role of oxygen. The results obtained by Plaksin and his school[42-45] clearly indicate that oxygen plays a very important role in the adsorption of xanthate. Plaksin[57] has put forth the view that oxygen promotes the dehydration of the surface of the mineral, thus facilitating the penetration of xanthate and its consequent fixation. Another function of oxygen is to mop up the free electrons present in the conduction band of galena which is known to be a semiconductor of n type and thus facilitate the adsorption of anionic xanthate. Oxygen functions as an electron acceptor and converts galena from an n-type to a p-type semiconductor. Plaksin and Shafeev[58] have found a parabolic relationship between the change in concentration of the electric charge carrier in galena and the oxygen concentration. The equation for the reduction of free electrons by the action of oxygen is given as:

$$\Delta n_e = - AC_{O_2}^{\frac{1}{2}} \qquad (7\ 6)$$

where n_e is the electric charge carrier concentration, C_{O_2} is the concentration of oxygen in the liquid phase of the pulp, and A and $\frac{1}{4}$ are proportionality coefficients.

ELECTROCHEMICAL STUDIES ON THE GALENA-XANTHATE SYSTEM

The role of oxygen in the adsorption of xanthate at a galena surface and the view put forth by Plaksin have been confirmed by Tolun and Kitchener[59] by electrochemical measurements. The effect of xanthate on the potential of a galena electrode was first studied by Lintern and Adam[60] in 1935; they found that when used as a cathode galena was little affected by xanthates, but adsorbed xanthate lowered the emf of anodic galena by about 0.2 V. The electrode potential of galena varies with pH, becoming more negative with an increase in pH indicating the adsorption of hydroxyl ions.[61] Subsequently Mukai and Wakamatsu[62] showed that xanthates form a reversible electrode system with mercury. Salamy and Nixon[63] demonstrated the electrochemical nature of the reaction between a mercury surface and xanthates using polarization curves. This method of investigation was applied to a galena-xanthate system by Tolun and Kitchener.[59] A clean lead electrode when examined in an air-free base solution (in a specially constructed all-glass cell) showed a static potential of -0.62 V which rose to -0.57 V on slight anodic polarization. The potential again fell to -0.70 V when the electrode was brought in contact with an air-free xanthate solution; this was shown to result from the formation of lead xanthate by the reaction:

$$Pb(OH)_2 + 2S = C\begin{matrix} \diagup SK \\ \diagdown OC_2H_5 \end{matrix} \longrightarrow Pb\left[\begin{matrix} S = C - S \\ \ \ | \\ \ \ OC_2H_5 \end{matrix}\right]_2 + 2KOH \qquad (7\text{-}7)$$

However, the surface remained hydrophilic as revealed by a nitrogen bubble that failed to stick. The bubble contact occurred only on further anodic polarization when the potential rose to +0.8 V. At this stage lead dioxide was formed and oxidized xanthate to dixanthogen by the reaction:

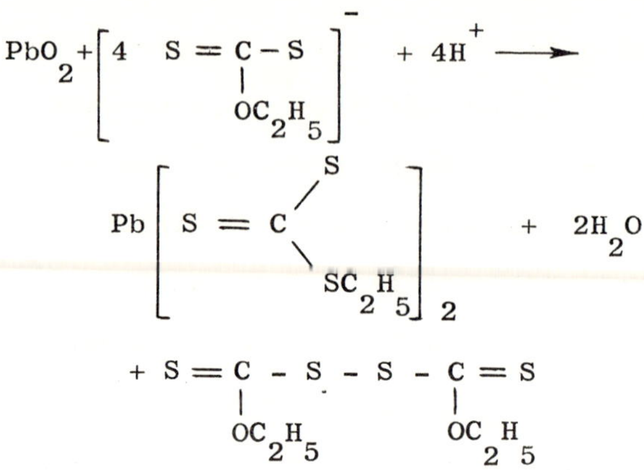

(7-8)

When the current was cut off, the potential fell rapidly (curve 2, Fig. 7-4) to -0.5 V, and the bubble contact failed after all the dixanthogen was reduced back to xanthate by metallic lead.

The polarograms obtained using a galena electrode are shown in Fig. 7-5. This electrode showed a potential of -0.28 V in a buffer solution. When brought in contact with a 10^{-3} M solution of potassium ethyl xanthate it fell to -0.31 V, but no bubble contact was observed. With an anodic current the potential rose to -0.13 V, which agrees with the redox potential for the xanthate-dixanthogen system, and at this stage the bubble contact developed.

The results of Tolun and Kitchener[59] conclusively demonstrate the dual role of oxygen in the

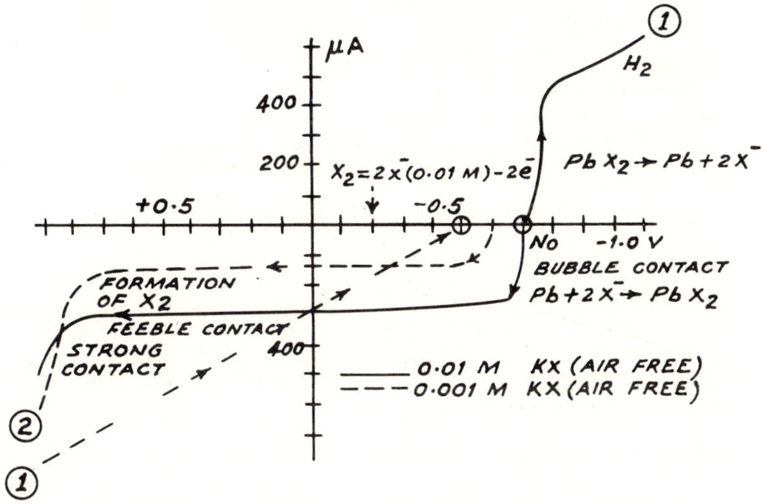

FIG. 7-4

Polarograms for potassium ethyl xanthate on lead. [After R. Tolun and J.A. Kitchener: Trans. Inst. Mining Met., 313-322, 1964; reproduced by permission of Institute of the Mining and Metallurgy, London.]

adsorption of xanthate by galena; oxygen first superficially oxidizes galena, creating a film of basic thiosulfate which interacts with xanthate, forming a lead xanthate film on the surface. Second, it raises the electrochemical potential of the galena until the xanthate begins to be oxidized to dixanthogen. Thus, a mixed film of xanthate and dixanthogen renders the mineral surface hydrophobic.

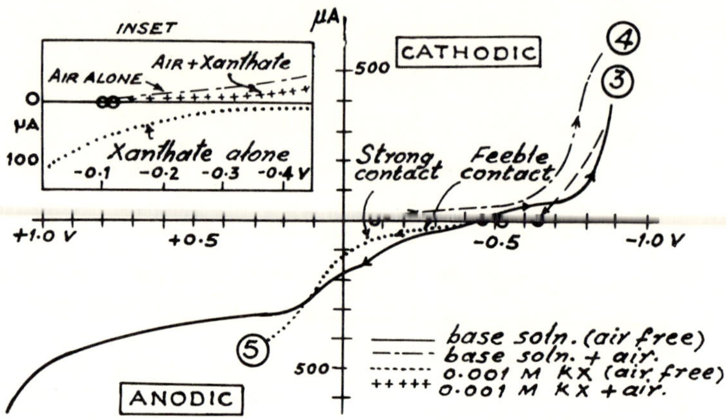

FIG. 7-5

Polarograms at a galena electrode. [After R. Tolun and J.A. Kitchener, Trans. Inst. Mining Met., 313-322 (1964); reproduced by permission of Institute of the Mining and Metallurgy, London.]

The electrochemical studies on a galena surface reported by Plaksin and Shafeyev[64] have revealed nonuniform adsorption of xanthate at different regions as a result of variation in the electrochemical potential. The adsorption measurements made by microradiographic tehcniques showed that xanthate fixation occurs in anodic potential regions, and when a crystal of galena was kept under voltage there was a rapid uptake of xanthate when the potential rose to positive values, which is in conformity with the findings of Tolun and Kitchener.[59]

INFRARED SPECTRA OF XANTHATE ADSORBED AT A GALENA SURFACE

Since 1962 investigators in Canada and the United States have employed infrared spectroscopy for the elucidation of the nature of the adsorbed species at a galena surface. Greenler[65] was the first to report the results of infrared studies employing the KBr pellet technique. Lead sulfide samples that had been exposed to air showed infrared evidence of lead carbonate, lead sulfate, lead sulfite, and other lead sulfur oxygen compounds. When such samples were treated with an aqueous solution of potassium ethyl xanthate, sulfoxylate ions were displaced; these results are in complete accord with the findings of earlier investigators mentioned above. Lead sulfide samples heated in air to produce more oxidation of the sulfide ions adsorbed more xanthate from solution than less oxidized samples. A surface xanthate compound was formed on a lead sulfide that had the infrared spectrum of lead ethyl xanthate. Part of the surface xanthate was readily removed by washing in acetone. The amount of xanthate remaining on the lead sulfide surface after repeated washing in acetone was approximately equivalent to a monolayer of xanthate ions. For this chemisorbed species Greenler[65] suggested the structure:

$$\begin{array}{ccccccccc} & & S & & & & Pb & & S \\ & & \| & & & & S & & Pb- \\ C_2H_5O & - & C & - & S & - & Pb & & S \\ & & & & & & S & & Pb \end{array}$$

in which xanthate ions are attached to lead atoms, which in turn are bound to sulfur atoms of the lead sulfide lattice; in normal lead ethyl xanthate

$$\begin{array}{ccccccccccccc} C_2H_5O & - & C & - & S & - & Pb & - & S & - & C & - & OC_2H_5 \\ & & \| & & & & & & & & \| & & \\ & & S & & & & & & & & S & & \end{array}$$

the sulfur atoms are attached to a xanthate ion. The infrared spectram of the two compounds would be expected to resemble each other closely.

These results have been largely confirmed by Leja and co-workers[66,67] who worked with the surface of natural galena. Evaporated lead sulfide substrates appeared to be more active toward adsorption of xanthate than the polished surface of natural galena, since acetone or benzene dissolved all the adsorbed lead ethyl xanthate from natural galena but not from evaporated lead sulfide films. The infrared spectra of adsorbed species are shown in Fig. 7-6, which also shows the spectrum of a freshly deposited lead sulfide film. When these spectra were compared with those of the ethyl xanthate compounds described in Chapter 6, the adsorbed species could be identified as multilayers of lead ethyl xanthate (curve b, Fig. 7-6). When these substrates were subjected to a high vacuum (10^{-6}-10^{-7} mm) for periods of a few days, most of the xanthate was desorbed from the surface; the action of a vacuum in desorbing the multilayers is comparable to the action of acetone or benzene, which dissolve the adsorbed multilayers. The remaining xanthate was found to be present in monolayer thickness and exhibited a definite shift in the high-frequency C-O-C band from about 1210 to 1195 cm^{-1}. In addition, a weak band appeared at about 1265 cm^{-1}, which was attributed to the formation of some dixanthogen on the surface (compare with the spectrum of dixanthogen in Chapter 6). This band was not found when lead sulfide films were washed in diethyl ether, which dissolved lead xanthate (present as multilayers) as well as dixanthogen.

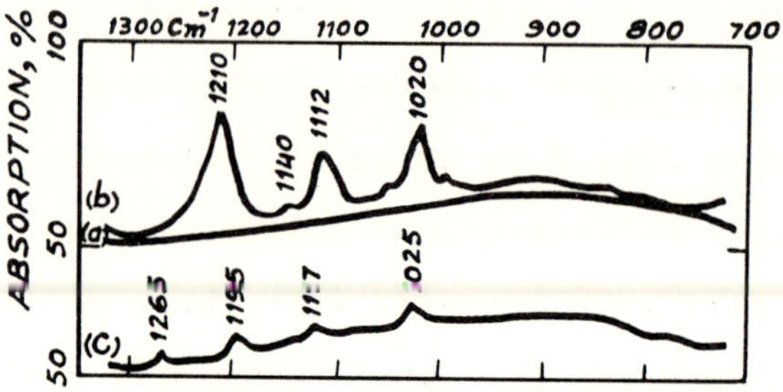

FIG. 7-6

Spectra of ethyl xanthate adsorbed on lead sulfide. (a) Freshly deposited Pb-S films; (b) adsorbed multilayers of lead ethyl xanthate; (c) = (b) after 2 days under high vacuum. [After G.W. Poling and J. Leja: J. Phys. Chem., 67, 2121 (1963); copyright 1963 by the American Chemical Society and reproduced by permission]

Another interesting observation of Poling and Leja[67] supports the role of oxygen in oxidizing xanthate and promoting adsorption of a mixed film of xanthate and dixanthogen. They found that treatment in deoxygenated solutions containing diethyl dixanthogen resulted in the adsorption of a near-monomolecular coating of lead ethyl xanthate. The high-frequency C-O-C band occurred at about 1195 cm^{-1}. This shift in C-O-C band frequency has been interpreted as indicating a 1:1 coordination between the surface metal atom and the xanthate radical in the first monolayer, that is, Pb-EtX.

In contrast, in multilayer coatings of lead xanthate and bulk lead ethyl xanthate precipitate, the lead and xanthate are bound by 1:2 coordination. This agrees with the findings of Hagihara,[40] who showed by electron diffraction that in the first xanthate monolayer lead and xanthate are bound by 1:1 coordination. The data strongly suggest that under normal conditions the dissolved oxygen oxidizes xanthate anions in solution to dixanthogen which is coadsorbed with xanthate; this is in accord with the conclusions of Tolun and Kitchener.[59] According to Poling and Leja,[67] this reaction would be expected. Since galena develops a negative surface potential on immersion in water,[58] dixanthogen, which is a neutral molecule, is adsorbed in preference to anionic xanthate and once it has passed through the electrical double layer it is partially reduced at the surface, forming a film of lead xanthate.

The infrared spectrum of a galena-xanthate system has been recently reexamined by Coleman and Powell,[68] who employed the attenuated reflection method in conjunction with standard procedures of infrared spectroscopy. In addition to confirming the findings of Greenler,[65] these authors have shown that the adsorption of ethyl xanthate on a galena surface is pH dependent and occurs in the pH region 7-11.

It is now appropriate to examine the contact angle data of Wark and Cox[29] and others in the light of the above discussion of the role of oxygen in the adsorption of xanthate at a galena surface. In their contact angle studies (pp. 254-6) Wark and Cox[29] also found that a galena surface treated with ethyl xanthate samples that had been kept in the atmosphere for a considerable length of time showed a contact angle of 80°, while the magnitude of the pure xanthate angle was 60°. The angle of 80° was identified with the formation of a dixanthogen film at the galena surface. Higher xanthates also form films of the corresponding dixanthogens in the presence of oxygen. The contact angle values for a series of xanthates and dixanthogens formed under oxidizing conditions were reported by Rao and Patel,[69] who found that in the absence of air or oxygen a mineral conditioned with ethyl xanthate showed an angle of 60°, thus confirming the observations of Wark and Cox.[29] The fact that ethyl xanthate alone gives rise to a contact angle of 60°, while with diethyl dixanthogen the angle rises to 80°, appears to indicate that xanthate alone could form an adsorbed film at a galena surface. This apparent disagreement between the classic work and the results of the recent investigations discussed above can be tentatively explained by suggesting that the value of 60° observed at a galena surface could be the characteristic contact angle of a mixed film of xanthate and dixanthogen, while the value of 80° is attributable to a pure dixanthogen film. Although no direct confirmation of this view has been attempted, it is appropriate here to admit that in the contact angle studies reported by previous workers no special precautions were taken to rigidly exclude oxygen from the system, and the presence of small quantities of dissolved oxygen could well give rise to a mixed xanthate-dixanthogen film. In support of this, in a recent contact angle study of a galena surface, Beebe and Westley[70] have shown that appreciable time is required for the interaction of galena with oxygen either before or during its contact with xanthate solutions.

The suggestion that a mixed xanthate-dixanthogen film is responsible for the creation of a hydrophobic film has not so far been supported by independent structural evidence. The principal difficulty has been in distinguishing between ionic xanthate and molecular dixanthogen in the adsorbed state by known experimental technique. Furthermore, it is hard to explain why the adsorption of xanthate ion alone, which is also known to possess the well-known heteropolar structure, should not create a hydrophobic surface and in what way the presence of dixanthogen alongside xanthate brings out hydrophobicity in the surface. In this context the recent work by Kakovsky and Arashkevich[71] on the collector action of dixanthogens as such is very illuminating. They have established the chemical nature of adsorption of dixanthogens at a galena surface. It was shown that a dixanthogen film is removed by treatment of the mineral with sodium sulfide in the same way xanthate ions are desorbed by this depressant. This indicates that the action of dixanthogen at the sulfide surface results in the formation of a chemically fixed layer of xanthate of the corresponding heavy metal. This is in agreement with the infrared spectroscopic evidence advanced by Leja and co-workers,[66] who observed the reduction of dixanthogen through the acceptance of electrons from the galena lattice. However, it is further mentioned by Kakovski and Arashkevich[71] that, although both xanthate ion and dixanthogen are fixed chemically, they show different degrees of resistance to the action of depressants of intermediate activity such as soluble chromates. In the presence of such depressants dixanthogen is a weaker collector than xanthate, since with dixanthogen there is an insufficient concentration of xanthate ion to counteract the depressant. If a xanthate is used as a collector, the concentration of the ion in the pulp may be sufficient to ensure the fixation of the collector ions on the mineral surface. If, however, dixanthogen is added to the pulp or formed by the oxidation of xanthate in the pulp, the xanthate ions in the solution would appear only as a result of the following consecutive reactions:

$$PbS + X_2 \rightleftharpoons PbX_2 + S^0 \qquad (7\text{-}9)$$

$$PbX_2 + 2OH^- \rightleftharpoons Pb(OH)_2 + 2X^- \qquad (7\text{-}10)$$

Hence, to obtain the same ion concentration in the latter case a complete reduction of dixanthogen is necessary, which requires a very long and practically impossible contact time between the mineral and the reagents. Therefore, the ionic form of the collector resists the action of depressants better than the molecular form, that is, dixanthogen.

It follows from this work that in the end xanthate ions are the adsorbed species bound to the galena lattice. This rules out the speculated mixed film of xanthate-dixanthogen. Also, it further supports the simple ion-exchange mechanism. A serious question arises in connection with contact angle values measured by Wark and Cox,[29] however. As mentioned before, contact angle of 80° has been believed to be characteristic of dixanthogen, as distinct from that of xanthate which gives an angle of 60°. The mechanism suggested by Kakovsky and Arashkevich,[71] however, requires that a sulfide mineral treated with dixanthogen give the same contact angle as one treated with xanthate. The discrepancy remains to be satisfactorily explained. The possibility of multilayers of dixanthogens giving rise to the higher contact angle appears worth considering.

It is thus seen from the foregoing discussion that xanthate is adsorbed at a galena surface by an ion-exchange mechanism. However, the adsorbed species is not identical with the bulk lead xanthate precipitate. Furthermore, the partial oxidation of xanthate to dixanthogen, giving rise to a mixed film, is a distinct possibility. No definite picture of the mixed xanthate-dixanthogen film has yet been found. However, by recent electron diffraction and crystal structure studies, Hagihara and associates[72] have arrived at a structural representation of a chemisorbed monomolecular film of xanthate. This is shown in Fig. 7-7. This representation essentially agrees with the molecular formula of Greenler mentioned above (p. 290) and postulates the existence of a lead xanthate molecule in which the lead atom shows a monovalent character in the vicinity of the crystal surface. This film is formed under conditions of such low concentration of xanthate in solution that the formation of lead xanthate precipitate is prevented. The lead atom of the Pb-SSCOR molecule

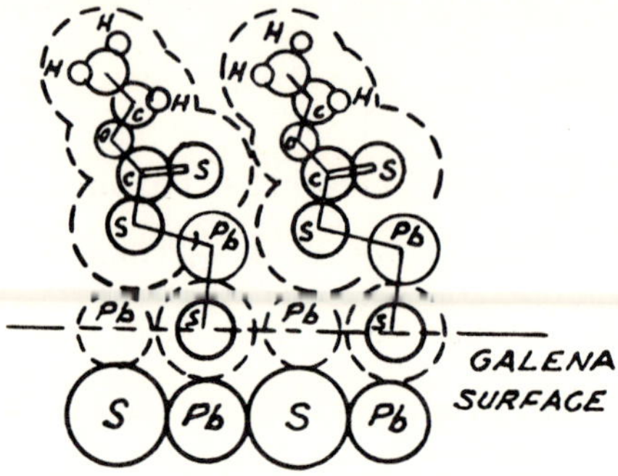

FIG. 7-7

presumably interacts with the sulfur atom of the galena lattice, regaining its divalent character. This monolayer adsorption is in accord with the ion-exchange mechanism. The surface lead ions are originally bound with hydroxyl or an equivalent number of oxysulfur ions and the latter are exchanged for xanthate ions. This also agrees with the enthalpy data of Mellgren discussed earlier.

THICKNESS AND STRENGTH OF XANTHATE ADSORBED IN FLOTATION PRACTICE

Although xanthates are known to be adsorbed in multilayers on galena when the mineral is left in contact with them, in actual flotation practice this multilayer is not significant. It has been shown by Gaudin and Preller[73] that an incomplete monolayer of xanthate suffices to produce a substantially complete recovery. It was shown that for

90% recovery of galena the surface coverage needed was only 63%; this was, moreover, an upper limit rather than an actual measurement, as the quantity of ion not adhering to galena was not calculated when the xanthate consumption was evaluated.

Hassialis and Myers,[74] who conducted air bubble contact experiments, inferred that adsorbed xanthate shows mobility at the mineral-solution interface, providing even distribution of the adsorbed xanthate. In our opinion such movement of the adsorbed ions could only be possible with molecules adsorbed by van der Waals forces and not those held by chemisorption bonds. The chemisorbed nature of the xanthate film has been amply demonstrated by the thermochemical, electrochemical, and infrared studies mentioned previously and has been further confirmed by Majumdar,[75] who found that xanthate film adsorbed on a galena surface is not destroyed by running tap water. Furthermore, an even distribution of the xanthate, as observed by Hassialis and Myers,[74] is in direct contrast to the observation of a nonuniform patched film reported by Plaksin and Shafeyev.[64] Much experimental work is needed before definite conclusion can be drawn regarding the mobility of the film.

Determination of the entropy of the adsorbed film would probably help in clarifying this aspect of the problem.

IRON PYRITE

Some of the early studies on pyrite (FeS_2) and xanthates revealed that the surface reactions are rather complex, which is not surprising in view of the variable oxidation state exhibited by iron in different compounds. Gaudin and Wilkinson[50] found that iron pyrite abstracted xanthate from its solution and that the adsorbed xanthate was oxidized to the corresponding dixanthogen. In a later investigation Gaudin and co-workers[30] found very unstable ferric xanthate at a pyrite surface when the latter reacted with xanthate. The ferric xanthate was found to undergo transformation, producing sulfur and an unidentified oil. The xanthate abstracted was not leached in benzene although the latter is known to be a good solvent for ferric xanthate. This was explained by Shvedov,[76] who showed that the solubility of the surface compound formed when pyrite abstracts xanthate is much lower than that of bulk ferric xanthate. Shvedov and Shorsher[77] determined the adsorption of potassium ethyl xanthate on iron pyrite. Their results are recorded in Table 7-7.

TABLE 7-7

Adsorption of Potassium Ethyl Xanthate on Pyrite

Quantity of xanthate added, mole/liter	Amount adsorbed on mineral: Percentage of that added	Amount adsorbed on mineral: Monomolecular layer, %
5.4×10^{-5}	60.7	239
23.2×10^{-5}	23.5	394
54.4×10^{-5}	11.6	457

The results of Table 7-7 show a continuous uptake of xanthate which is similar to the adsorption of xanthate by galena. This observation was reexamined by Gaudin et al.,[78] who determined the adsorption density of xanthate by employing a radiotracer technique (see Chapter 5). It was found that the continuous uptake of xanthate far exceeding that required for a monolayer film was caused by the action of oxygen; for when oxygen was rigidly excluded from the system, the maximum adsorption density corresponded very nearly to a monolayer and there was no further uptake. Clearly oxygen plays an important role. Gaudin and co-workers also found that the effluent solution in the adsorption experiments contained minute oily droplets, while no evidence of turbidity was seen in the influent solutions. The oily substance in the effluent was presumably dixanthogen formed by the oxidation of xanthate catalyzed by the pyrite surface.

The formation of dixanthogen in xanthate adsorption by pyrite has been supported by the electrochemical studies of Mitrofanov and Ryskin,[79] and more recently the Majima and Takeda.[80] Double-layer capacity on an electrode of a single crystal of pyrite was measured as a function of potential value. A pyrite electrode in a potassium sulfate solution has a characteristic capacity minimum at a potential of 0.15 V (with reference to a calomel electrode). In a solution of potassium sulfate containing 10^{-4} M butyl xanthate, the capacity minimum shifts to 2.0 V, and butyl xanthate is adsorbed on the surface of the pyrite at a potential exceeding -0.4 V, as indicated by a sharp drop in the capacity value in this region. As this is close to the oxidation potential of xanthate (though not identical with it), it was inferred that dixanthogen is the species adsorbed at the pyrite surface. In the cathode region, dixanthogen is reduced to xanthate, which is desorbed from the surface of pyrite at potential values of -0.5 to -0.55 V. This is indicated by a well-defined maximum in the capacity of the pyrite electrode.

Fuerstenau and co-workers[81] have made Nernst E.M.F. measurements on pyrite-xanthate systems and concluded that the species responsible for the formation of a hydrophobic film when xanthate is added as a collector is dixanthogen. Measured oxidation potentials of this system show that xanthate should be oxidized to dixanthogen, the presence of which is confirmed by infrared absorption. The authors have also calculated the concentration ratio [dixanthogen] / $[\text{xanthate ion}]^2$ at various pH values. The values so obtained show that at pH 9 and below almost all of the added xanthate exists in the form of dixanthogen, and that at pH 11 and above the amount of dixanthogen is exceedingly small. The depression of pyrite at pH 11 and above is attributed to the lack of dixanthogen. The depression at low pH values is attributable to the decomposition of xanthic acid, but the depression of pyrite around pH 6 cannot be the result of a lack of dixanthogen. No satisfactory explanation has been advanced for this anomaly. The adsorption of dixanthogen has been explained by suggesting the resonance bonding of the pyrite sulfur with the sulfur of the dixanthogen molecule. Independent structural evidence would be of interest in deciding the role of dixanthogen in this adsorption mechanism.

Mukai and associates[82] have shown that the adsorption of ethyl xanthate on iron pyrite is expressed by the Freundlich isotherm:

$$\mu = a \left[X^-\right]^{1/n} \tag{7-11}$$

where μ is the amount adsorbed, $[X^-]$ is the equilibrium concentration of ethyl xanthate and $1/n$ and a are constants. The adsorbability of xanthate is considered on the basis of the exponent $1/n$. Mukai et al. also estimated by thermodynamic calculation the standard free energy change for the adsorption reaction. The chemical affinity between ethyl xanthate and the sulfide minerals agrees well with the chemical affinity between sulfide ion and ferric ion. The authors infer that the xanthate ion combines with the surface of pyrite through the sulfur atom. No suggestion regarding the configuration of the adsorbed species has been made. An infrared spectroscopic study would probably provide better insight.

CHALCOCITE

The first set of experiments to elucidate the reaction between potassium xanthate and chalcocite were conducted by Dewey.[30] The mineral abstracted ethyl xanthate, and a number of compounds containing Cu_2S and cuprous xanthate in different molar ratios were identified by leaching the surface species in solvents such as methanol, acetone, ether, and benzene. Even after leaching with these solvents, the mineral remained highly floatable, thus indicating that the collector film consists of a species that is unleachable in any of the solvents employed. Schuhmann and Gaudin[38] confirmed these observations and further showed that the quantity of leachable xanthate increases with an increase in the initial concentration of xanthate that is used reaching a maximum of about 0.145 meq per 150 g of chalcocite; this value is shown to correspond to a monolayer film of xanthate. These results show that unleachable xanthate is formed even when xanthate is added in a starvation quantities and that it corresponds to the stable chemisorbed species formed by the exchange adsorption of xanthate ions at a chalcocite surface. The leachable xanthate is of no significance in the flotation and represents the "wasted" xanthate that occurs as cuprous xanthate of varying composition formed by the double decomposition of the oxidized mineral surface, containing cupric ions and hydroxyl ions, and potassium xanthate according to the following mechanism:

$$Cu_2S^{A} + X^- \longrightarrow Cu_2S^{X^-} + A^- \qquad (7\text{-}12)$$

(Unleachable Xanthate)

$$XCu_2S\ Y\ CuS + \tfrac{1}{2}O_2 + H_2O \longrightarrow$$

$$(X - 2)Cu_2S(Y + 2)\ CuS + 2Cu^+ + 2OH^- \qquad (7\text{-}13)$$

$$2Cu^+ + 2X^- \longrightarrow Cu_2X_2 \quad \text{(Leachable entity)} \qquad (7\text{-}14)$$

The detailed structure of the adsorbed species, the "unleachable entity" of Schumann's experiments has not been investigated. It is interesting to note that this unleachable xanthate can nevertheless be dissolved by pyridine, which is known to form charge-transfer complexes. Surprisingly, no further studies have been carried out.

SPHALERITE

The adsorption of ethyl xanthate on sphalerite follows the activation of mineral by cupric sulfate. The unactivated mineral fails to adsorb ethyl xanthate unless the pH is lowered to 6.6. This is attributed to the fair solubility of zinc ethyl xanthate. Shvedov and Andreeva[83] have, however, reported that sphalerite (presumably unactivated) adsorb xanthate, the uptake being of the same order of magnitude as that observed with pyrite by Shvedov and Shorsher.[77] It has been shown by calculation that the anions of ethyl xanthate transfer to the surface of sphalerite from an aqueous solution of xanthate of a concentration 100 times less than the value of the solubility of zinc ethyl xanthate, which is nearly 4.4×10^{-3} mole/liter. Interesting as these results appear to be, they do not seem to have found support from investigators in other laboratories. It has been repeatedly shown that, at least with ethyl xanthate, the sphalerite surface must be activated by cupric ions before satisfactory flotation can be achieved.

The fact that sphalerite cannot be floated by ethyl xanthate has been explained by the evidence advanced from crystal structure studies of zinc ethyl xanthate (see Chapter 6). In this molecule zinc atoms are bonded to the sulfur atoms of the xanthate groups by tetrahedral covalent bonds, forming a two-dimensional network of atoms. Such a bonding property hinders the adsorption of each isolated xanthic cation, or zinc xanthate molecule or molecular ion, to the atom sites of the sphalerite surface and, accordingly, will be inactive in developing the hydrophobic property on it. When the concentration of zinc anions in the solution is increased, zinc xanthate crystallites are precipitated; however, the side planes of the thick platelike crystallites are bounded by the methyl ends of the alkyl groups, and they lack the capacity for strong adhesion to the mineral surface.

The adsorption of xanthates at a activated sphalerite surface was first studied by Sato[84,85] by employing electron diffraction. The diffraction pattern showed that at the sphalerite surface activated by cupric ions curprous xanthate was formed. There were two kinds of crystallites of this compound; those with random orientation and those with a fiber orientation with the fiber axis normal to the cleavage face. It was inferred that in the fibrous structure the molecules of cuprous xanthate are built up with an end-to-end arrangement.

The investigations of Yonezawa[86] have shown that the adsorption of potassium ethyl xanthate at a sphalerite surface at pH 6.6 follows the Freundlich isotherm. The saturation adsorption density varies with pH and is given by the equation

$$\log \mu = \log \mu^{o} - k(\mathrm{pH}) \qquad (7\text{-}15)$$

where μ^{o} is the limiting value of μ, the saturation adsorption at pH 0, and k is a constant depending on the concentration of xanthate ion. The adsorption isotherm is of the form:

$$[H^{+}]^{0.15} \quad [X^{-}] = m \qquad (7\text{-}16)$$

where $[X^-]$ denotes xanthate ion concentration and m is a constant. The rate of adsorption obeys the law

$$\Gamma = \Gamma_1 + k \log t \qquad (7\text{-}17)$$

where Γ_1 is the adsorption at t = 1 min and k is related to the initial concentration c of ethyl xanthate by k = 10 + 0.4c. The rate of adsorption decreases as the pH of the solution increases. It shows a positive temperature coefficient from which it is inferred that some chemical reaction is taking place. Yonezawa also determined the adsorption of ethyl xanthate at a sphalerite surface activated by cupric sulfate. The adsorption follows a S-shaped curve and rises from 4 to 55 mg per gram of solid as the concentration of cupric sulfate increases from 10^{-6} to 10^{-2} moles/liter. The surface coverage by cupric ions and by xanthate ions corresponds roughly to a monolayer at the cupric ion concentration 1.4×10^{-5} mole/liter, at which the molar ratio of the two becomes 2.3 and is equal to the ratio of the cross-sectional areas of cuprous ion and xanthate ion. These values are in accord with the finding of Sato that the xanthate is adsorbed as cuprous ethyl xanthate.

The desorption of xanthate from a sphalerite surface can be effected by sodium sulfide, sodium cyanide, or ferric sulfate as shown by Yonezawa.[86] In the case of the first two, the desorption results from the displacement of xanthate by the competing anions. The mechanism of desorption by ferric sulfate has not been established.

The collector mechanism of xanthate at a zinc sulfide surface has recently been studied by Yamasaki and Usui[87] by infrared spectroscopic examination. The results show a striking similarity to the collector mechanism operative at a galena surface. Infrared spectra have revealed that the nature of the adsorbed species is dependent on the extent of oxidation of the zinc sulfide sample. On an unoxidized sample the amount of adsorbed xanthate does not exceed the amount of xanthate calculated for unimolecular adsorption. The infrared bands indicate that the xanthate adsorbed on an unoxidized sample is subject to an inhomogenous electric field of the lattice. However, a sample of zinc sulfide that had been subjected to superficial oxidation by hydrogen peroxide and then treated with xanthate showed infrared adsorption bands similar to those of zinc dodecyl xanthate.

The analysis of products in the suspension remaining after adsorption of xanthate has revealed that xanthate ion is adsorbed by an ion-exchange mechanism, an equivalent quantity of an anion such as sulfate being released in the process. In the case of an oxidized sample, in addition to the monolayer of xanthate, zinc dodecyl xanthate is formed by a reaction between potassium dodecyl xanthate and the oxidation product of the mineral.

Yamasaki and Usui have further attempted to identify the adsorbed species by leaching it in pyridine and recording the infrared spectrum of the product obtained on adding water to the pyridine extract. The spectrum so obtained with the oxidized sample is identical with that of zinc dodecyl xanthate, as is to be expected. The pyridine extract containing the unoxidized sample did not give a precipitate on adding water. The infrared spectrum of the pyridine extract shows an adsorption band that corresponds to the presence of carbon disulfide. This was further confirmed by adding potassium ethylate to the pyridine extract; potassium ethyl xanthate was formed and was identified by its characteristic absorption peak at 3010 Å. From this it is concluded that the monolayer is composed of dodecyl xanthic acid or of ion pairs consisting of dodecyl xanthate ions adsorbed on the zinc atoms and hydrogen ions adsorbed on the sulfur atoms of the zinc sulfide surface. This is also in accord with the instability of the adsorbed film on an unoxidized sample, since on exposure to the atomosphere the H^+ ions and xanthate ions would unite, forming unstable xanthic acid. This "ion pair" adsorption mechanism tends to support the theory of Cook and Nixon (p. 263).

Recently Steininger[88] has produced further support in favor of Cook and Nixon's theory. It is significant that his analysis also concerns a collector mechanism for xanthates on sphalerite surfaces. His experimental results indicate that flotation is prevented when about 95 % of the collector is ionized. This has been interpreted on the basis of the neutral molecule adsorption mechanism and is represented by the two steps

$$-Zn^+ + H_2O \longrightarrow -ZnOH + H^+ \qquad (7\text{-}18)$$

where $-Zn^+$ represents the zinc surface site.

$$-ZnOH + RSH \longrightarrow -ZnSR + H_2O \qquad (7\text{-}19)$$

where R stands for an alkyl group and -ZnSR represents the collector adsorbed on the mineral surface. It has been shown that the upper pH limit of flotation is directly and closely related to the pK_a of the neutral molecule (xanthic acid formed by the hydrolysis of xanthate) collector. Also,

$$pH = pK_a + \log \frac{[RS^-]}{[RSH]} \tag{7-20}$$

and at constant collector addition C. For low flotation recoveries when the amount of collector adsorbed on the mineral is small compared to C,

$$[RS^-] \simeq C - [RSH] \tag{7-21}$$

It follows, therefore, that

$$pH \simeq pK_a + \log \left[\frac{C}{RSH} - 1 \right] \tag{7-22}$$

According to the mechanism proposed, the surface density of the adsorbed collector varies with [RSH] and with the density of hydroxylated surface sites [-ZnSR]. Hence.

$$[-ZnSR] = Ks\,[RSH]\,[-ZnOH] \tag{7-23}$$

where Ks is a constant, or

$$[RSH] = \frac{[-ZnSR]}{Ks[-ZnOH]} \qquad (7\text{-}24)$$

It can be assumed that flotation recovery is determined by the surface density of the adsorbed collector $[-ZnSR]$ at low surface coverage. This is expressed as

$$\rho = f\,[C\text{-}ZnSR] \text{ or } [-ZnSR] = g\,\rho \qquad (7\text{-}25)$$

and hence, it follows that

$$[RSH] = \frac{g\,\rho}{Ks[-ZnOH]} \qquad (7\text{-}26)$$

$$\text{and } pH \simeq pK_a + \log\left[\frac{C\ Ks\ [-ZnOH]}{g\,\rho}\right]^{-1} \qquad (7\text{-}27)$$

From the last expression, it follows that for a constant density of hydroxylated surface sites and at constant flotation recovery there is a linear relationship between pH and pK_a if Ks is the same for different collectors. For constant zero recovery, pH becomes the upper critical pH, pH_c This has been confirmed by flotation tests. The expected linear relation between pH_c and pK_a has been shown over the pH range 6-12 for five different sulfdydryl collectors.

The mechanism suggested above explains the well-known fact that sphalerite is not effectively floated by short-chain xanthates such as potassium ethyl xanthate. Since the pK_a value decreases with a decreasing hydrocarbon chain, the lower hydrocarbon chain xanthates are strongly ionized, and since the zinc salts of these xanthates are relatively soluble, a hydrophobic film is not formed on a sphalerite surface. With un-ionized collectors, the formation of water also contributes to the free energy of the reaction and represents a supplementary driving force for the replacement of OH groups by collector entities.

The analysis of Steininger does not take into consideration the stability of the collector film at a sphalerite surface in relation to the stability of the neutral molecule in solution. It is well known that xanthic acids are unstable in water, their half-life being of the order of a few minutes. If this applies to the collector film at the mineral surface also, flotation by neutral molecule adsorption would be rather untenable. It is possible, however, to explain the mechanism by postulating that the adsorbed state of the collector is not identical with the bulk state. Furthermore, Yamasaki and Usui[87] have preferred ion pair adsorption to neutral molecule adsorption. In a recent publication Cook[39] has pointed out that while the ion pair adsorption mechanism and neutral molecule adsorption mechanism are superficially alike, free energy considerations favor the ion pair mechanism. This would necessitate the adsorption of xanthate ions together with H^+ ion. Thus, with regard to sphalerite, the collector mechanism is still rather ambiguous. A thermochemical study of the system might well reveal more information in this direction.

MARMATITE

This mineral is actually sphalerite containing varying amounts of FeS in the lattice. The adsorption of ethyl xanthate at this mineral surface was studied by Bushell[7], who expressed the xanthate uptake by the expression:

$$S = K' [m]^n \ [H^+]^q \qquad (7\text{-}28)$$

where m denotes the equilibrium concentration of xanthate and the constants n and q were shown to have the values of 0.774 and 0.437, respectively. The relation is similar to the one found for the adsorption of xanthate at a galena surface. Although marmatite contains iron in the lattice, it is sometimes necessary to activate the mineral by cupric ions for the satisfactory adsorption of xanthate. Rao and Patel[89] have shown by contact angle measurements that marmatite activated with cupric ions shows an angle of 60° when conditioned with ethyl xanthate, while the unactivated mineral shows an angle of only 45°.

CHALCOPYRITE

This is a very common mineral of copper, but surprisingly few fundamental studies have been carried out on it. The only work so far known concerns the effect of oxygen on the adsorption of xanthate on chalcopyrite reported by Plaksin and co-workers.[43] The results (Fig. 7-8) show the significant role of oxygen in enhancing the adsorption of xanthate and, consequently, flotation of the mineral. It is seen that with a very small concentration of oxygen stable chemisorption of xanthate on the mineral does not occur. It may be recalled that the similar influence of oxygen on the adsorption of xanthate at a galena surface has been extensively investigated by Plaksin and co-workers[42,44,45] and substantiated by others.[55,59] The basic question

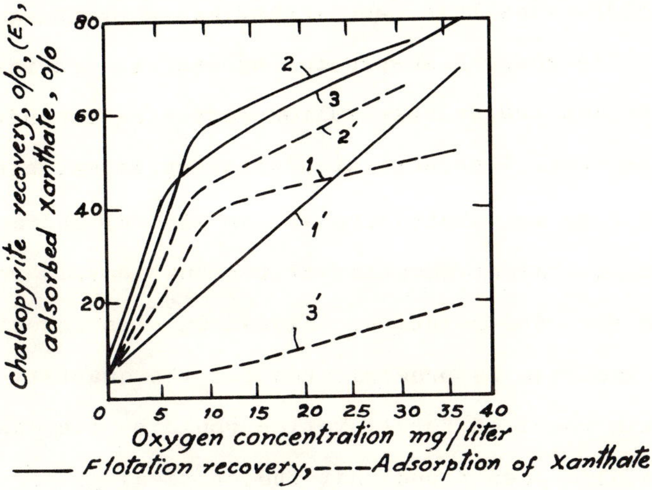

FIG. 7-8

Influence of oxygen on the flotation of chalcopyrite from quartz and on the adsorption of potassium ethyl xanthate by chalcopyrite. (1 and 1') Addition of xanthate, 10 g/ton; (2 and 2') addition of xanthate, 100 g/ton; (3 and 3') addition of xanthate 1000 g/ton. [After I.N. Plaksin, C.V. Bessonov and V.I. Tiurnikova, Proc. Acad. Sci. USSR, 102, No.2 (1955).]

whether or not oxygen is essential for the adsorption of xanthate has not been investigated in depth for minerals other than galena.

PYRRHOTITE (FeS)

The flotation of this mineral by xanthate also follows activation by cupric ions. The flotation characteristics have been investigated by Chang et al.[9] In aqueous suspension pyrrhotite liberates ferrous ion and soluble sulfoxylate ions including sulfate ion. When activated with cupric sulfate, cupric ions are abstracted at the mineral surface and an equivalent quantity of ferrous ions is released into the solution. Excess Cu^{2+} is undesirable, since xanthate is precipitated and the quantity of xanthate required for flotation would be increased. It has also been found that the critical pH for pyrrhotite is 4.5 and that amyl xanthate is superior to ethyl xanthate for collecting pyrrhotite. With amyl xanthate nearly 100% recovery is obtained with only 3 lb per ton of collector consumption, while with ethyl xanthate only up to 80% mineral is recovered by using 6 lb per ton of collector.

The adsorption of xanthate at a pyrrhotite surface, investigated by Bushell,[7] follows an expression similar to Eq. (7-28) above. It is similar to the ones determined for the adsorption of xanthate at galena and marmatite surfaces. The mechanism of adsorption has so far not been elucidated, but it is likely that it follows a pattern similar to that observed for sphalerite activated with Cu^{2+}.

MOLYBDENITE

In industrial practice xanthates are not commonly used for benefication of this mineral. However, the adsorption of xanthate by molybdenite which has been measured by radiotracer techniques[90] should be of special interest to surface chemists since molybdenite is known to be nonpolar, and the mechanism of adsorption of anionic xanthate would probably throw more light on the surface reactions of xanthates. It is likely that the oxidation product dixanthogen, which is a nonpolar molecule, plays a major role in the flotation of molybdenite.

OTHER SULFIDE MINERALS

Some preliminary work on the adsorption of xanthates by stibnite (Sb_2S_3), cinnabar (HgS), and realgar (As_2S_3) has been reported by Glembotskii and associates[8]. These minerals were subjected to sorption and desorption tests by the use of radioactive isotopes. The adsorption of xanthate on these minerals is promoted by Cu^{2+}, Pb^{2+} and Hg^{2+} introduced as activators. Desorption is achieved by potassium cyanide and alkalies.

PHYSICAL STABILITY OF THE ADSORBED XANTHATE FILM

Although the chemisorbed nature of xanthate at sulfide mineral surfaces is a well-known fact, the strength of the bond between the mineral lattice and the xanthate differs for individual minerals. This results in a variation in the physical stability of the adsorbed xanthate. Studies on the effect of thermal treatment on the xanthate film performed by Glembotskii[91] have shown that the xanthate adsorbed at a chalcopyrite surface is the easiest to destroy; treatment with water at 100°C for 2 min was enough to remove 99.5% of the adsorbed xanthate. Within the same time and by the same treatment, the amounts of adsorbed xanthate removed from sphalerite and pyrite were 10 and 30% respectively. Glembotskii applied this variation in the thermal stability of xanthates adsorbed at different mineral surfaces in the selective destruction of adsorption layers of xanthate collectors, thus facilitating selective separation of the individual minerals. Tests were conducted with galena, sphalerite, pyrite, and chalcopyrite using ethyl xanthate (50 g/ton for galena and 60 g/ton for other minerals), while for the flotation of sphalerite 300 g/ton of cupric sulfate was added as an activator. The temperature was raised to 100°C by passing an electric current through the mineral layer for 1 or 2 min for galena, pyrite, and chalcopyrite, and for 6 min for sphalerite because the adsorption layers on sphalerite had been found to be very stable. The weights of the mineral particles that passed into the tailings upon flotation after passage of the current were 32%, 50%, 95%, and 97% for galena, sphalerite, pyrite, and chalcopyrite, respectively.

The thermal stability of the adsorption layers formed by amyl xanthate was found to be lower than that of ethyl xanthate. It is likely that oxidation of the adsorbed film, enhanced by the action of heat, is also an important factor determining its stability. It has been shown by Rao and Patel[69] that adsorbed amyl xanthate is more easily oxidized than ethyl xanthate, and it is possible that the bond strength of the resultant dixanthogen film is weaker than that of the corresponding xanthate. In support of this view it has been shown that the xanthate film adsorbed at a sphalerite surface does not undergo appreciable oxidation, as revealed by contact angle measurements,[92] and this agrees with Glembotskii's observation that the xanthate film at a sphalerite surface is most resistant to thermal effects. A thermochemical study on the bond strengths of the adsorbed xanthate and dixanthogen films is likely to prove fruitful.

7. XANTHATES IN MINERAL FLOTATION

EFFECT OF REGULATING AGENTS ON THE ADSORPTION OF XANTHATES AND FLOTATION RECOVERY

In industrial practice regulating agents are often used to accomplish selective flotation of minerals. The action of such reagents depends upon their ability to prevent the adsorption of xanthates or to promote the adsorption of xanthates. Mention has already been made of the latter class of modifying agents, commonly called activators, in the xanthate flotation of sphalerite, pyrrhotite, and nonsulfide minerals such as malachite, chrysocolla, and quartz. The other class of regulating agents known as depressants prevent the adsorption of xanthates, and the optimum concentration varies for different minerals, thus facilitating the selective separation of minerals.

Foremost among the depressing agents are hydroxyl ions; the optimum concentration of hydroxyl ions, expressed in terms of pH, required to prevent the adsorption of xanthate has already been noted. In addition to hydroxyl ions, cyanide and sulfide ions are also adsorbed effectively at sulfide mineral surfaces and prevent the adsorption of xanthate anions. Wark and Cox[93] studied the action of sulfide on the adsorption of xanthates; their results are summarized in Table 7-8. Values for the concentration of hydrosulfide were calculated using the two dissociation constants for H_2S.

TABLE 7-8

Concentration of Hydrosulfide Ion Required to Prevent the Collecting Action of Potassium Ethyl Xanthate Added in the Amount of 25 mg (equivalent to 56 μmoles/liter)[a]

Mineral	HS^-, mg/liter	HS^-, μmoles/liter
Galena	0.007	0.25
Chalcopyrite	0.30	9
Bornite	1.3	40
Covellite	1.7	50
Pyrite	2.5	75
Chalcocite	6.4	190

[a] From M.A. Cook and G.A. Last: J. Phys. Chem., 56, 637-642 (1952).

It is seen that hydrosulfide ion has a higher mineral surface activity than xanthate ion for all the minerals tested except chalcocite, the ratio of potency between depressant ion and collector ion being as high as about 600:1 for galena. It may also be noted by a comparison of Tables 7-8 and 7-4 that the increase in concentration of hydrosulfide ions needed to prevent the adsorption of xanthate approximately parallels the increase in critical pH value of the mineral.

Further studies to establish the relationship between the concentration of xanthate ion and that of hydrosulfide ion were conducted by Cook and Last.[94] Their data have been presented in a more general fashion by Gaudin[4j] (see Fig. 7-9). Except for the line for pH 5.7, all the straight lines relating the concentration of xanthate ion with the concentration of hydrosulfide ion coincide, which suggests that the ratio $[HS^-]/[X^-]$ is independent of pH. The data also confirm that hydrosulfide ion is approximately 10 times more potent than the amyl xanthate ion with respect to surface activity on galena. Furthermore, since the hydroxyl ion concentration at pH 13.3 is 200 mmoles/liter, and

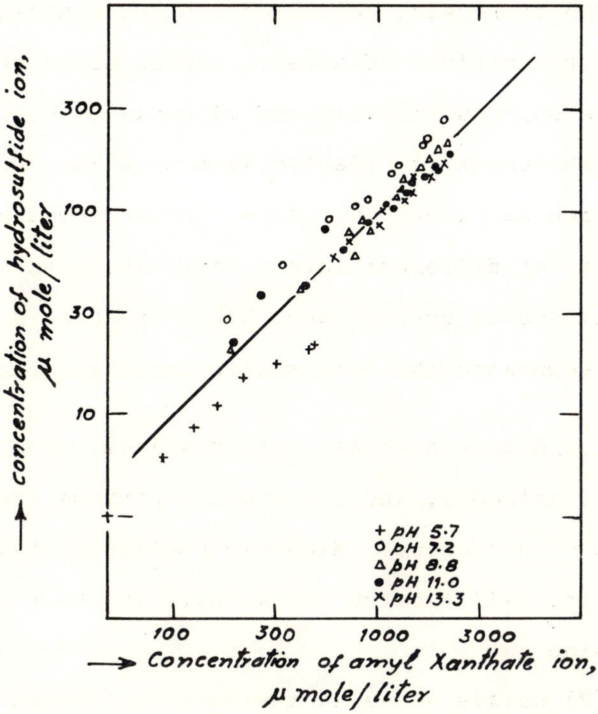

FIG. 7-9

Relationship between hydrosulfide and amyl xanthate ion for critical conditions in the flotation of galena. [Data of G.A. Last,[94] from A.M. Gaudin, "Flotation," McGraw Hill, New York, 1957, p. 292; and reproduced by permission.]

equivalent depression is obtainable with 60 μmoles of hydrosulfide ion per liter, the latter may be regarded as 3000 times as effective a depressant as hydroxyl ion.

The depressing action of cyanide also follows similar principles. Kakovsky[95] has inferred that cyanides depress flotation of minerals by solubilizing xanthates, thus effectively preventing their adsorption at mineral surfaces. He showed that xanthates of different metals have widely varying solubilities in cyanide and, based on these data, he has separated the metals into three groups:

(1) Metals such as lead, thallium, bismuth, arsenic, antimony, and tin whose xanthates are insoluble in cyanide. Minerals containing these metals are well floated by xanthate in the presence of cyanide.

(2) Metals whose xanthates show increased solubilities in the presence of cyanide. Such metals are platinum, mercury, silver, cadmium, and copper. Flotation of minerals containing these metals is depressed at high cyanide ion concentrations.

(3) Metals whose xanthates are readily soluble in cyanide solutions; iron, gold, nickel, palladium, and zinc belong to this class. Minerals of these metals are easily depressed by cyanide ion.

The solubility of metal xanthates in cyanide also depends on the hydrocarbon chain of xanthates, solubility decreasing with increasing length of hydrocarbon chain. Consequently, the depressant action of cyanide becomes less effective with increasing chain length. Because of this, the selectivity of the depressing action of cyanide in flotation is decreased with xanthates having longer hydrocarbon chains.

From the above considerations it is evident that the efficiency of separation of any two minerals is determined by the difference in the solubilities in cyanide of the xanthates of their corresponding metals. This is confirmed in industrial practice of flotation of sulfide ores by xanthates in the presence of cyanide. As noted above, lead and zinc differ considerably in the solubilities of their xanthates in cyanide, and this permits the selective flotation of galena from sphalerite which is preferentially depressed by cyanide.

ADSORPTION OF XANTHATES ON NONSULFIDE MINERALS

The adsorption of xanthates has been studied in several nonsulfide minerals, notably cerussite ($PbCO_3$), malachite ($CuCO_3$), quartz (SiO_2), and chrysocolla (copper silicate), and in elements such as graphite, gold, copper, and nickel. In the case of nonsulfide minerals the adsorption of xanthate is sometimes facilitated by pretreatment of the mineral with sodium sulfide. Shorsher[96] investigated the adsorption of ethyl and amyl xanthates on cerussite and showed that the tenacity of xanthate adsorption is very much lower than that on sulfide minerals such as galena and chalcocite. Xanthates adsorbed on carbonate minerals are easily removed from the surface by treatment of the minerals with organic solvents. This was also shown experimentally for amyl xanthate adsorbed on malachite.

The uptake of xanthate by cerussite has been measured,[4k] and it is inferred that xanthate ion from the solution is precipitated in the form of lead xanthate and that carbonate ions are released into the solution:

$$PbCO_3 + 2S{=}\underset{\displaystyle OR}{\underset{|}{C}}{-}S^- \longrightarrow Pb\left[S{=}\underset{\displaystyle OR}{\underset{|}{C}}{-}S\right]_2 + CO_3^{2-} \qquad (7\text{-}29)$$

According to this mechanism, the adsorption of xanthate is dependent on the solubility product of the lead xanthate formed, and this value continues to decrease with decreasing length of hydrocarbon chain. The nature of the adsorbed species is still speculative since it is not certain whether or not carbonate ion is also adsorbed on lead xanthate and gives rise to a patchwork of lead ions, xanthate ions, and carbonate ions.

Fleming[32] has determined the relationship between pH and concentration of potassium ethyl xanthate necessary to induce contact at a cerussite surface and the relationship between pH and the concentration of sodium carbonate necessary to prevent contact using 100 mg/liter of xanthate. It was shown that the lead ion concentration at this concentration of xanthate was 1.7×10^{-9} moles/liter, which was very close to the lead ion concentration present in the aqueous carbonate solution that contained the critical carbonate ion concentration. Carbonate ion is thus the controlling factor in determining the critical pH of cerussite (see also pp. 256-258).

The adsorption of xanthate at cerussite and malachite is greater when the mineral has been pretreated with sodium sulfide. Rehbinder[97] found that the activating action of sodium sulfide is particularly strong in the presence of a short-chain xanthate. In order to obtain a contact angle of 90° on fresh malachite it was necessary to treat its surface in an aqueous solution containing 1500×10^{-5} mole/liter of ethyl xanthate, while the sulfidized mineral showed the same contact angle in a solution containing only 17×10^{-5} mole/liter of xanthate, i.e., nearly 100 times less than in the first case. Mitrofanov and co-workers[98] showed that preliminary sulfidization of cerussite not only increased the amount of xanthate adsorbed but also increased the rate of uptake. However, excessive sulfide ion concentrations (above 6 gram equivalents per ton of ore) diminish the uptake because of the desorption of xanthate by sulfide ions.

The adsorption of xanthate by chrysocolla (copper silicate) pretreated with sodium sulfide was described by Wright and Prosser.[19] By infrared spectroscopic examination they identified cuprous xanthate and dixanthogen formed by the reaction:

$$Cu^{2+} + 2EtX^{-} \longrightarrow Cu(EtX)_2 \qquad (7\text{-}30)$$

$$2Cu(EtX)_2 \longrightarrow 2CuEtX + (EtX)_2 \quad (7\text{-}31)$$

The distribution of copper xanthate and dixanthogen at different pH values was analyzed; Table 7-9 records some of the results of Wright and Prosser which are explained as follows.

The copper dispersed in the solution (column 5) was determined by polarographic methods and accounts for both cuprous ethyl xanthate and abraded chrysocolla. Cuprous ethyl xanthate was not dissolved in the solution, therefore, copper compounds must have been derived from abraded chrysocolla particles. Comparison of columns 3 and 5 show that 80-90% of the cuprous ethyl xanthate was dispersed. Furthermore, approximately 55% of the total diethyl dixanthogen was detected in the solution and there is no mass balance when the values in column 4 are compared with the sum of values in columns 6 and 7. This is probably attributable to the fact that the reagent (isooctane) used to dissolve dixanthogen was immiscible in the solution in the mineral pores, and thus the dixanthogen trapped inside the pores could not be detected.

TABLE 7-9

Distribution of Copper Compounds and Dixanthogen after Reacting Chrysocolla with Ethyl Xanthate

pH	Reaction time, min	Total product		Quantity in filtrate		$(EtX)_2$ at
		CuEtX	$(EtX)_2$	Cu^{2+}	$(EtX)_2$	Surface
5.5	15	7.6	3.8	6.2	2.1	0.22
10.0	10	0.75	0.38	0.67	0.20	0.02
8.0	30	2.30	1.15	—	0.65	0.10

The results show that most of the products of reaction, although formed at the surface, are subsequently dispersed in the solution and only about 5% are available at the surface to facilitate the flotation of the mineral. The reason for this is mainly physical; the products of reaction do not adhere to the chrysocolla surface.

Adsorption of xanthate on quartz preactivated with lead ions has been described by Fuerstenau and co-workers.[99] The best flotation recovery was observed in the pH range 5.8-8.5 in the presence of 1 X 10^{-4} M Pb^{2+} and 1 X 10^{-4} M xanthate. In this concentration region very little xanthate would be present. The mechanism suggested by the authors is as follows:

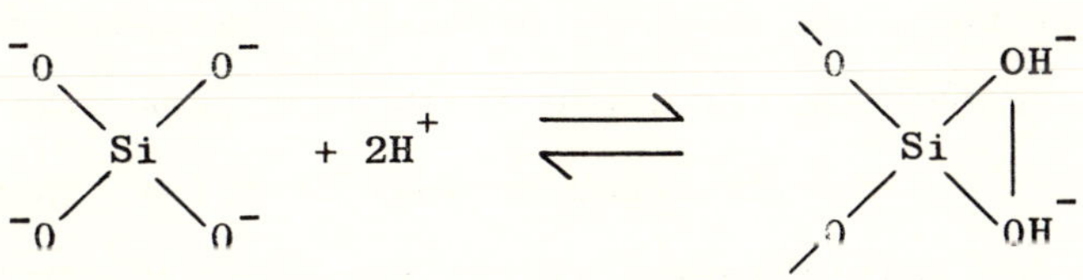

(7-32)

(Quartz in water)

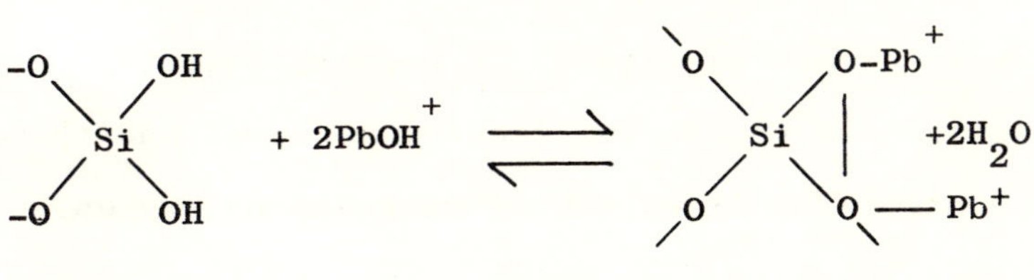

(Activation step)

This reaction is followed by the adsorption of xanthate ion at the positive surface sites:

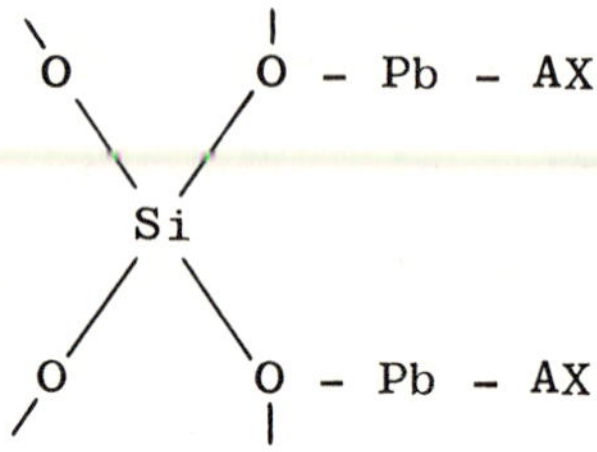

where AX denotes an amyl xanthate radical.

The adsorption of xanthates on nonpolar minerals such as coal or graphite has attracted some attention in recent years in view of the fact that the collection of these minerals by xanthates has been shown in laboratory tests[24,25] although they would not be expected to adsorb xanthate ions because of their nonpolar nature. Horsley and associates[100] have shown by contact angle tests that pure alkali xanthates are not adsorbed on coal, but that they are effectively adsorbed in the presence of oxygen. Oxidation results in the formation of dixanthogen, which is a powerful collector for coals of all ranks except lignites. It appears that the carbon surface serves as an electron acceptor and thus promotes the oxidation of xanthate by oxygen. The dixanthogen formed, being nonpolar, is adsorbed on the nonpolar mineral surface; this has been confirmed by Wallum and Mellgren,[101] who observed that the consumption of xanthate increased linearly with the graphite surface area in the presence of air or pure oxygen. The consumption was greater in the presence of air than in the presence of oxygen, but this was probably connected with the lowering of the pH by the dissolved carbon dioxide.

Leja and co-workers[66,67,102] have extended infrared spectroscopic studies to the adsorption of xanthate and dixanthogen on elemental nickel and copper. The object was to investigate the configuration of the adsorbed species on a metal atom. Infrared spectra of the nickel alkyl xanthates (ethyl, butyl, and n-amyl) were compared with the spectra of the corresponding potassium alkyl xanthates and dixanthogens, and with the spectra obtained when these compounds were adsorbed on the metals. Adsorption of xanthates from acidic aqueous solutions of nickel plates produced multilayer coatings of nickel xanthates. Adsorption of diethyl dixanthogen vapor at nearly 10^{-12} mm of mercury on oxidized or sulfidized surfaces resulted in the catalytic dissociation of dixanthogen to form adsorbed nickel ethyl xanthate with coadsorption of excess dixanthogen. The surfaces obtained by either of the above adsorption treatments were strongly hydrophobic, but the adsorbed layers were readily removed by washing with an organic solvent, prolonged soaking in water, or evacuation at 10^{-3} mm of mercury and at a temperature above $35^{\circ}C$. With metallic copper surfaces, both oxidized as well as sulfidized, it was shown that cuprous ethyl xanthate was the main constituent of the adsorbed layers when the adsorption of xanthate or dixanthogen was studied by infrared spectroscopy. When dixanthogen alone was used, it dissociated on adsorption to form cuprous ethyl xanthate. Dixanthogen molecules present in a partially oxidized xanthate solution or in a dixanthogen emulsion were also coadsorbed, unaltered chemically, on the initially coadsorbed cuprous xanthate. This rendered the surface more hydrophobic than when cuprous xanthate alone was adsorbed. The phenomenon of mixed adsorption of xanthate and dixanthogen was similar to the one observed with galena, as discussed above.

The effect of temperature on the rate of xanthate uptake at copper powder surfaces[103] showed that with reduced copper, temperature has no appreciable effect on the rate of abstraction of xanthate, but with an oxidized copper surface the rate of xanthate ion abstraction decreases with rising temperature. This is attributed to the increasing solubility of the overlying multilayer mixture of cuprous xanthate and dixanthogen (cf. pp. 290-291). The magnitude of the activation energy as calculated by the Arrhenius plot was 22.6 kcal/mole, suggesting chemisorption. The entropy of activation was found to have a small negative value, indicating that the degree of ordering in the adsorbed layer is increased. This work on metallic surfaces is of special interest in the formation of more definite conclusions regarding the nature of the surface reactions of xanthates and the nature of the adsorbed species.

REACTION OF XANTHATE IN THE AIR-LIQUID PHASE IN THE FLOTATION SYSTEM

In the mineral flotation system, aside from the collector, the frother plays a major role in stabilizing air bubbles by lowering the surface tension of water. As both collector and frother are surface active, the interaction between the two at the air-water interface where the frother molecules concentrate was considered by some workers as one of the important factors in flotation. Bowcott[104] first studied the reactions of the xanthate ion at the air-water interface by surface pressure measurements on a monolayer film of cetyl alcohol penetrated by potassium lauryl xanthate. There was a rapid rise in the observed surface pressure from 3.7 dynes/cm up to a maximum of 50 dynes/cm, which was reached with a xanthate concentration of 3×10^{-6} M in the underlying solution. This indicated an interaction of xanthate with alcohol at the air-water interface. After the initial rise the surface pressure decreased slowly. Similar observations were made on monolayer films of a long-chain carboxylic acid. In this case the film pressure rose on injecting potassium lauryl xanthate but decayed rapidly, showing a weak interaction between xanthate and carboxylate, both of which are anionic. The rapid fall in pressure is believed to be a result of the oxidation of xanthate. With cetylamine monolayer films the trend was similar to that observed with lauryl alcohol films, but with a fully N-substituted amine such as cetyltrimethylammonium iodide the association of long-chain xanthate was extremely strong and stable. The force-area curve of an equimolecular mixed film of cetyltrimethylammonium bromide and potassium ethyl xanthate showed that the mixed film remains stable up to 67 dynes/cm, at which stage xanthate would be in a highly condensed state and as such would be very stable against atmospheric oxidation. However, the film collapsed on injecting iodine beneath the surface because of the rapid oxidation of xanthate. The stable association between xanthate and the quaternary ammonium salt is attributed to the strong interaction between oppositely charged ions.

Buckenham and Schulman[105,106] have used surface tension measurements to establish the existence of a molecular association between xanthates and cetyltrimethylammonium bromide. Their results support the conclusions of Bowcott[104] on the strong association formed when xanthates and N-substituted amines are combined and give rise to molecular complexes in which the ratio of xanthate to amine is 1:1. The resulting molecular complexes are effective flotation collectors for sulfide minerals and also for quartz. It was found, however, that the collectors adsorbed at a quartz surface could easily be washed out, while the adsorbed xanthate film at a sphalerite surface remains stable on washing and the mineral can be effectively floated by adding octyltrimethylammonium iodide. This difference in behavior of the two minerals could be taken advantage of in the effective separation of quartz and sphalerite by first conditioning and then washing the mineral. This action of xanthate in the presence of a cationic system has lead to the conclusion that at a sphalerite surface xanthate is adsorbed over the positive metal ion. This will be followed by a strong association between xanthate and quaternary ammonium ions as shown by monolayer and surface tension studies. The latter become attached to the adsorbed xanthate, resulting in increased hydrophobicity. A similar mechanism involving the adsorption of the ammonium ion over the negative sites of quartz followed by interaction and association with xanthate ion is suggested as an explanation of the collecting properties of xanthate mixed with cetyltrimethylammonium bromide on quartz. The molecular association at the mineral surface is represented in Fig. 7-10.

Pomianowski and Leja[107] measured the adsorption of equimolar solutions of xanthate and alkyl

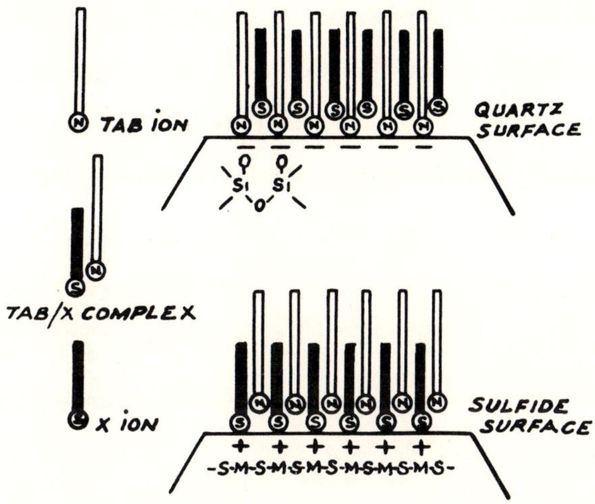

FIG. 7-10

Schematic representation of the attachment of xanthate/alkyltrimethylammonium bromide collectors. [After M.H. Buckenham and J.H. Schulman: Trans. AIME, 226, 1 (1963); reproduced by permission of the AIME.]

trimethylammonium bromide on copper, nickel, and sphalerite powders. In the aqueous solution there was no indication of molecular association complex formation since no shift in the utraviolet absorption bands and no new bands were observed. However, the opalescene and precipitation effects occurring within the mixtures of higher homologs of xanthate and long-chain alkyl trimethylammonium bromide strongly indicated that some interaction did occur.

This association appeared to be mostly electrostatic in nature as previously suggested by Bowcott[104] and as such could not be detected by ultraviolet spectra. It was subsequently detected by X-ray diffraction patterns which showed that in no case did spacings of the precipitate formed by a 1:1 molar ratio of xanthate and trialkylammonium bromide agree with the d spacings of the constituents, suggesting that a new compound was formed. The presence of alkyl trimethylammonium bromide strongly increased the preferential abstraction and the rate of abstraction of ethyl xanthate by copper and nickel, but there was no significant influence in sphalerite on which a near monolayer of xanthate was formed. There appears to be some disagreement with the results of Buckenham and Schulman,[105] who found increased hydrophobicity on a sphalerite surface when a mixture of xanthate and alkyl trimethylammonium bromide was used. In more recent publications Leja and co-workers[103,108,109] have mentioned that the apparent preferential adsorption of xanthate species on copper surfaces indicates that the 1:1 complex between xanthate and alkyl trimethylammonium bromide is not permanently adsorbed on the surface, although it plays a major role in increasing the preferential uptake of xanthate species. After diffusing through the electrical double layer at the solid-liquid interface, it is readily decomposed and allows the alkyl trimethylammonium ion either to diffuse back into the bulk solution or to act as a link in the transfer of a further amount of xanthate anion by serving as a proton donor to or electron acceptor from xanthate ions. This mechanism presumes that adsorption of xanthate when it is present alone is hindered by the negatively charged electrical double layer at copper surface. This was in fact experimentally shown by Guarnaschelli and Leja[103]; they found that when oxygen was rigidly excluded from the system freshly reduced copper powder did not adsorb xanthate, but xanthate uptake was observed when it was complexed with alkyl trimethylammonium bromide. Their results demonstrate the possibility of adsorption of complexed xanthate under anaerobic or deoxygenated conditions. It is now widely accepted that xanthate ions cannot adsorb at sulfide and metallic surfaces in the absence of oxygen to effect partial conversion of xanthate to dixanthogen. With regard to the use of alkyl trimethylammonium halides in combination with xanthates, Finkelstein[110] has very recently pointed out that these long-chain quaternary ammonium salts catalyze the oxidation of xanthate to dixanthogen, thus promoting the collector action of xanthates. This possibility has to be considered along with the collector mechanism under anaerobic conditions mentioned previously. There seems to be some disagreement between the results of Finkelstein and those of others on the action of quarternary ammonium salts. This may be mainly attributable to variations in experimental techniques employed. Further work is desirable. A comparative infrared study of the adsorbed film of complexed xanthate and the mixed xanthate-dixanthogen film is likely to be of great interest in further elucidating the mechanism of xanthate adsorption.

REFERENCES

1. Keller, C.H.: U.S. Patent 1,554,216 (1925).

2. Taggart, A.F. "Elements of Ore Dressing," Wiley, New York, 1951, Chapt. 13 and 14, pp.234-263.

3. Pryor, E.J. "Mineral Processing," Mining Publications, Ltd, London, 1960, Chapt. 18, pp.510-560, Chapt. 20, pp.595-670.

4. Gaudin, A.M. "Flotation." McGraw Hill, New York, 1957 (a) pp.436-446; (b) pp.433-434;(c) p.452; (d) pp.431-433;(e) p.254;(f) p.251;(g) pp.253-255; (h) p.436;(i) pp.147-148;(j)pp. 291-293; (k) pp.251-253.

5. Berry, J.G.: Extraction of copper at Moubhandar Works. Indian Minerals, 4,113-124 (1950).

6. Rao, S.R. and C.C. Patel: Flotation of chalcopyrite by xanthates under oxidizing conditions and by dixanthogen. Trans. AIME, 226, 243-248 (1963).

7. Bushell, C.H.G.: Behaviour of xanthates in flotation. Can. Mining Met. Bull., 61, 65-77 (1958).

8. Glembotskii, V.A., P.M. Solozhenkin, L.L. Ognewa: Investigation of the interaction of reagents with the surface of stibnite, cinnabar and realgar. Izv. Akad. Nauk Tadzh. SSR Otd. Fiz. Takhn. i Khim Nauk, 64-78 (1965); Chem. Abstr. 64, 12253 (1966).

9. Chang, C.S., S.R.B. Cooke, and I. Iwasaki: Flotation characteristics of pyrrhotite with xanthates. Mining Eng. 6, 209-217 (1954).

10. Chuev, Y.A.: Use of xanthates for dressing of molybdenum. Obogashch. Rud, 5, No.2, 9-10 (1960); Chem. Abstr. 55, 25, 648 (1966).

11. Blaskett, K.S.: Flotation of copper-lead-cobalt-nickel ore from Rum Jungle, N.T. Commonwealth Scientific and Industrial Research Organization and the University of Melbourne: Ore Dressing Investigations, Rept. No.509 (1955).

12. Blaskett, K.S.: Flotation testing of a base metal ore from White's orebody, Rum Jungle, N.T. J. Mines, Metals Fuels, 8, No.7, 33-40, 46 (1960).

13. Gaudin, A.M. and A. E. Anderson: Flotation Fundamentals, V. Utah Eng. Exptl. Station Tech. Paper No. 9 (1930).

14. Johnson R.A.: The reaction of certain sulphur-bearing collectors with cerussite and galena. Thesis, Montana School of Mines Butte, Montana (1933).

15. Fleming, M.G.: Effects of soluble sulfide in the flotation of secondary lead minerals. Paper 24, Symposium on Mineral Dressing, Institution of Mining and Metallurgy, London, September, 1952.

16. Rey, M. and P. Raffinot: The flotation of oxidized zinc ores. Paper 30, Symposium on Mineral Dressing, Institution of Mining and Metallurgy, London, September, 1952.

17. Rey, M. and V. Formanek: Some of the factors affecting selectivity in the differential flotation of lead-zinc ores particularly in the presence of oxidized lead minerals. Paper 18, Group III, International Mineral Processing Congress; Institution of Mining and Metallurgy, London, April, 1960.

18. Rey, M., G. Sitia, P. Raffinot and V. Formanek: Flotation of oxidized zinc ores. Mining Eng., 6, 416-420 (1954).

19. Wright, A.J. and A.P. Prosser: Study of the reactions and flotation of chrysocolla with alkali metal xanthates and sulphides. Trans. Inst. Mining Metallurgy, 74, Pt. 5, 259-279, (1964).

20. Fahrenwald, A.W.: Xanthate and pine oil float native copper in amygdaloid ores. Eng. Mining J. 126, 58-59 (1928).

21. Benedict, C. Harry: "Lake Superior Milling Practice," Michigan College of Mining and Technology Press, Houghton, Michigan, 1955.

22. Das Gupta, S.B. and P.I.A. Narayanan: Recovery of sulphur and gold from the tailings of Nundydroog Mines, K.G.F. J. Sci. Ind. Res. 16A, 374 (1957).

23. Plaksin, I.N. and A.G. Lopatin: Effect of alkali on floatability of native gold. Izv. Vysshykh Uchebn. Zavedenii Tsvetn. Met., 3, (No.3), 38-44 (1960); Chem. Abstr. 55, 5205 (1961).

24. Majumdar, K.K.: Beneficiation of graphite from Mysore State. J. Sci. Ind. Res. (India), 7B, 167 (1948).

25. Janta, Milos: Froth flotation of Czechoslovak fine coals. Prace Ustavu Vyzkum Vyuziti Paliv Publ., 10, Nos. 30-42, 17-53 (1959); Chem. Abstr., 54, 12538 (1960).

26. Kakovskii, A.I.; E.I. Silina, and A.N. Grebneva: Areas of application of flotation reagents with high activity. Izv. Vysshykh Uchebn. Zavednii, Tsvetn. Met., 5, No.3, 42-48 (1962); Chem Abstr., 57, 14769 (1962).

27. French Patent, 1,152,380, February 14 (1958), Flotation of ore by long chain xanthates; Chem. Abstr., 54, 10, 791, (1960).

28. Anfimova, E.A., V.N. Gauke, A.A. Egizarov, G.A. Kepp, and D.M. Adorova: Laboratory and industrial experiments on the effect of combined xanthates on flotation of copper ore. "Intensifikatsiya Flotation Protsessa"; Akad. Nauk SSSR Gos. Kom. Po. Toplivn. Prom. Pri Gosplane SSR Inst. Gorn.Dela, pp.27-33 (1964); Chem. Abstr., 62, 11, 458 (1965).

29. Wark, I.W. and A.B. Cox: An experimental study of the effect of xanthates on contact angels at mineral surfaces. AIME, Tech. Publ. No. 461 (1932); Trans. AIME, 112, 189-244 (1934).

30. Gaudin, A.M., F. Dewey, W.E. Duncan, R.A. Johnson, and O.F. Tangel: Reactions of xanthates with sulphide minerals. Trans. AIME, 112, 319-347 (1934).

31. Arashkevich, V.M. and I.A. Kakovskii: Collecting properties of dixanthogens. Tsvetn. Metal., 35, No.5, 20-24 (1962).

32. Fleming, M.G.: Effects of alkalinity on the flotation of lead minerals. Trans. AIME, 193, 1231-1236 (1952).

33. Taggart, A.F., G.R.M. del Giudice, and O.A. Ziehl: The case for the chemical theory of flotation. Trans. AIME, 112, 348-381 (1934).

34. Wark, I.W., and A.B. Cox: Principles of Flotation I, Trans. AIME, 112, 189-244 (1934).

35. Cook, M.A. and J.C. Nixon: Theory of water-repellent films on solids formed by adsorption from aqueous solutions of heteropolar compounds. J. Phys. Colloid Chem., 54, 445-459 (1950).

36. Cox, A.B. and I.W. Wark: Nature of the adsorption of soluble collectors: J. Phys. Chem., 37, 797-803 (1933).

37. Wark, I.W. and A.B. Cox: Flotation of graphite and sulphur by collectors of xanthate type and its bearing on the theory of flotation. J. Phys. Chem., 39, 551-556 (1935).

38. Gaudin, A.M. and R. Schuhmann, Jr: The action of potassium n-amyl xanthate on chalcocite. J. Phys. Chem., 40, 257-275 (1936).

39. Cook, M.A.: Hydrolyticand ion-pair adsorption models for collectors in flotation. Trans. AIME., 223, 444-447 (1962).

40. Hagihara H.: Mono and multilayer adsorption of aqueous xanthate on galena surface. J. Phys. Chem., 56, 610-621 (1952).

41. Bhrany, U.N. and M.R.A. Rao: Contact angle studies at the pyrolusite surface. J.Sci. Ind. Res. (India), 12B, 590-596 (1953).

42. Plaksin, I.N.: The action of gases and reagents on minerals in the flotation process. Bull. Acad.Sci USSR; Tech. Sci. Sect. No. 12 (1950).

43. Bessonv, C.V., I.N. Plaksin, and V.I. Tiurnikova: The effect of various gases on the flotatibility of chalcopyrite. Bull.Acad.Sci. USSR., Tech. Sci.Sect. No.10 (1955).

44. Plaksin, I.N. and C.V. Bessonov: Adsorption of oxygen in sulphide suspensions. Proc.Acad. Sci. USSR., 98, No.2(1954).

45. Plaksin, I.N., C.V. Bessonov, and V.I. Tiurnikova: The mechanism of interaction of xanthates with surfaces of sulphide minerals. Proc.Acad.Sci. USSR., 102, No.2 (1955).

46. Taylor, T.C. and A.F. Knoll: Action of alkali xanthate on galena. Trans. AIME, 112, 382-397 (1934).

47. Herd, H.H. and W. Ure: Surface chemistry in the flotation of galena. J. Phys.Chem., 45, 93-106 (1941).

48. Knoll, A. and D.L. Baker: Adsorption of potassium xanthate by galena in oxygen-free atmosphere. AIME, Tech. Publ. 1313 (1941).

49. Taggart, A.F. and M.D. Hassialis: Solubility product and bubble attachment in flotation. AIME, Tech. Publ. 2078 (1946).

50. Gaudin, A.M. and W.D. Wilkinson: Surface action of some sulphur bearing organic compounds on some finelly ground sulphide minerals. J. Phys. Chem., 37, 833—845 (1933).

51. Mellgren, O.: Heat of adsorption and surface reaction of potassium ethyl xanthate on galena. Trans. AIME., 235, 46-60 (1966).

52. Bushell, C.H.G.: On the action of xanthate in flotation. Can.Mining Met. Bull., 68, 327-331 (1965).

53. Burkin, A.R., M.J. Rogers, and G.J.E. Hazel: Adsorption and reaction of xanthate at galena-liquid interfaces. In "VII International Mineral Processing Congress," Vol. 1, ed. N. Arbiter. Gordon and Breach, London, 1965, p. 337.

54. Mellgren, O. and M.G. Subha Rau: Adsorption of ethyl xanthate on galena. Trans. Inst. Mining Met., 72, Pt. 6, 425-442 (1962-1963).

55. Gaudin, A.M. and N.P. Finkelstein: Interactions in the system galena-potassium ethyl xanthate-oxygen. Nature, 207, 389-391 (1965).

56. Schulman, J.H. In "Recent Developments in Mineral Dressing" Inst.Mining and Metallurgy, London, 1953, pp.423-424.

57. Plaksin, I.N. and S.V. Bessonov: Role of gases in flotation systems. In "Proceedings of the Second International Congress on Surface Activity," Vol. III, ed. J.H. Schulman. Butterworth, London, 1957 pp.361-367.

58. Paksin, I.N. and R. Sh. Shafeev: Influence of surface properties of sulphide minerals on adsorption of flotation reagents. Trans. Inst. Mining Met., 72, 715 (1962-63).

59. Tolun, R. and J.A. Kitchener: Electrochemical study of the galena-xanthate-oxygen system. Trans. Inst. Mining Met., 73, 313-322 (1963-64).

60. Lintern, P.A. and N.K. Adam: The influence of adsorbed films on the potential difference between solids and aqueous solutions with special reference to the effect of xanthate on galena. Trans. Faraday Soc., 31, 564-574 (1935).

61. Kamienski, B. and A. Pomianowski: The influence of hydrogen ions on the potential of a mineral electrode during the process of flotation. Bull. Acad. Polon. Sci. Classe III, 2, 85 (1954).

62. Mukai, S. and T. Wakamtsu: The electrochemical study of flotation. Mem. Fac. Eng. Kyoto, 24, No.4 389-410 (1962).

63. Salamy, S.G. and J.C. Nixon: The application of electrochemical methods to flotation research. In "Recent Developments in Mineral Dressing" Institution of Mining and Metallurgy, London, 1953, pp.503-516.

64. Plaksin, I.N. and R.Sh.Shafeyev: Properties of distribution of xanthates on the surface of sulphide minerals. J. Mines, Metals,Fuels, 8, No.7, 1-8 (1960).

65. Greenler, R.C.: An infrared investigation of xanthate adsorption by lead sulphide. J. Phys. Chem., 66, 879 (1962).

66. Leja, J., L.H. Little, and G.W. Poling: Xanthate adsorption studies using infrared spectroscopy. Trans. Inst. Mining Met., 72, 414-423 (1962-1963).

67. Poling, G.W. and J. Leja: Infrared study of xanthate adsorption on vacuum deposited films of lead sulphide and metallic copper under conditions of controlled oxidation. J. Phys. Chem., 67, 2121-2126 (1963).

68. Coleman, R.E. and H.E. Powell: Infrared spectoscopy studies of a xanthate galena system. U.S. Bur. Mines, Rept. Invest. No.6816,24 pp. (1966).

69. Rao, S.R. and C.C. Patel: Behaviour of adsorbed xanthates at galena surface in the presence of oxygen and carbon dioxide. J. Sci. Ind. Res. (India), 21D, 8-10 (1962).

70. Beebe, R.R. and C.E. Westley: Contact angle on galena as a function of oxygen concentration. Trans. AIME, 220, 9-10 (1961).

71. Kakovsky, I.A. and V.M. Arashkevich: The study of the properties of organic disulphides. Paper No. S-8, VIII International Mineral Processing Congress, Leningrad (1968).

72. Hagihara, H., T. Sakurai, and T. Ikeda: Action of xanthate on sulphide mineral surfaces. In "Proceedings of the Fourth International Congress on Surface Active Substances, Brussels, 1964", Vol. II, ed. J. Th.G.Overbeek. Gordon and Breach, New York, 1967, pp. 369-371.

73. Gaudin, A.M. and G.S. Preller: Surface area of flotation concentrates and the thickness of collector coatings. AIME, Mining Technol. 10(3) Tech.Publ.2002 (1946).

74. Hassialis, M.D. and C.G. Myers: Collector mobility and bubble contact. Mining Eng., 3, 961-968 (1951).

75. Majumdar, K.K.: Stability of collector films on mineral surfaces. J. Sci. Ind. Res. (India), 11B, 392 (1952).

76. Shvedov, D.A.: Hypothesis to explain good flotability of oxides. Mineral Dressing. J. (Moscow), No.6 (1936).

77. Shvedov, D.A. and I.N. Shorsher: On the reaction of pyrite with xanthates. Scientific Investigatory Works on the theory and Practice of Flotation (Papers), No.2, ONTI (1938).

78. Gaudin, A.M., P.L. de Bruyn and O. Mellgren: On the adsorption of ethyl xanthate on pyrite. Mining Eng., 8, 65-70 (1956).

79. Mitrofanov, S.I. and M. Ya.Ryskin: Electrochemical studies on sulphide flotation. Paper No.S-5, VIII International Mineral Processing Congr. Leningrad (1968).

80. Majima, H. and M. Takeda: Electrochemical studies on the xanthate dixanthogen systems on pyrite. Trans. AIME, 241, 431 (1968).

81. Fuerstenau, M.C., M.C. Kuhn, and D.A. Elgillani: The role of dixanthogen in the xanthate flotation of pyrite. Trans. AIME, 241, 148 (1968).

82. Mukai, Shigeru, G. Kano, and M. Maruoka: Adsorption reaction between ethyl xanthate and sulphide minerals. Syiyokaishi, 15(1), 23-28 (1963).

83. Shvedov, D.A. and A.I. Andreeva: Effect of oxidation on flotation of sphererite. Scientific Investigatory Work on the Theory and Practice of Flotation (Papers), No.4, ONTI (1938).

84. Sato, R.: Basic study of flotation of sphalerite. J. Mining Inst. Japan, 70, 63-71, 415-419 (1954).

85. Sato, R.: Electron diffraction investigation of action of aqueous heavy metal salts on sulphide mineral cleavage faces; a basic study of activation in flotation. In "Proceedings of the Second International Congress on Surface Activity," Vol. III, ed. J.H. Schulman Butterworth, London, 1957, p.349.

86. Yonesawa, T.: Experimental study of adsorption and desorption of xanthate by sphalerite. Trans. Inst. Mining Met., 70, 329 (1960-1961).

87. Yamasaki, T. and S. Usui: Infrared spectroscopic studies of xanthate adsorbed on zinc sulphide. Trans. AIME, 232, 36 (1965).

88. Steininger, J.: Collector ionization in sphalerite flotation with sulfhydryl compounds. Trans. AIME, 238, 251 (1967).

89. Rao, S.R. and C.C. Patel: Behaviour of adsorbed xanthate at marmatite surface. Indian J. Tech., 1, 132-133 (1963).

90. Khan, G.A., A.G. Shpineva, and I.A. Fed'Kovskii: Xanthate adsorption by molybdenite. Tsvetn. Metal., 5, No.6, 29-34 (1962).

91. Glembotskii, V.A.: Physical stability of collector adsorption layers on mineral surfaces and methods for their destruction. International Mineral Processing Cong., 1960, pp.235-253; Institution of Mining and Metallurgy, London.

92. Rao, S.R. and C.C. Patel: Behaviour of adsorbed xanthates at sphalerite surface. J. Sci. Ind. Res. (India), 20D, 391-393 (1961).

93. Wark, I.W. and Cox, A.B.: An experimental study of influence of sodium sulphides, alkalies and copper sulphate on effect of xanthates at mineral surfaces. Trans. AIME., 13A, 7-25 (1939).

94. Cook, M.A. and G.A. Last: Collector depressant equilibrium in flotation. J. Phys. Chem., 56, 637-642 (1952).

95. Kakovsky, I.A.: A theory on the flotation action of cyanide. Proc. 2nd Sci-Tech. Sess. Makhanobr Inst; Metallurgizdat (1952).

96. Shorsher, I.N.: Flotation of carbonate minerals (cerussite and malachite). Gostoptekhizdat (1949).

97. Rehbinder, P.A.: "The Physical Chemistry of the Flotation Process." GNTI (1938).

98. Mitrofanov, S.I., V.G. Kushnikova, I.A. Strigin, and G.S. Rozhavski: An investigation of the sulphidization of oxide and carbonate minerals. Concentration Metallurgy of Non-ferrous metals and methods of Analysis (papers); Metallurgizdat (1955).

99. Fuerstenau, M.C.; J.D. Miller, R.E. Pray and B.F. Perinne: Metal ion activation in xanthate flotation of quartz. Trans. AIME, 232, 359-364 (1965).

100. Horsley, R.M.; H. El-Sinbawy, and H.G. Smith: Xanthate in coal flotation. Fuel, 31, 302 (1952).

101. Wallum, P.F. and O. Mellgren: private communication (1967)

102. Poling, G.W. and J. Leja: Infrared spectroscopy studies of xanthates and their adsorption on nickel. Can. Met. Quart., 1, No.2, 109-128 (1962).

103. Guarnaschelli, C. and J. Leja: Complexes formed by quarternary ammonium salts with xanthates and synergistic effects in adsorption of xanthate ions on reduced copper powders. Separation Sci., 1 (4), 413-429 (1966).

104. Bowcott, J.E.L.: Reactions of the xanthate ion at the air-water interface. In "Proceedings of the Second International Congress." Surface Activity, Vol. III, ed. J.H. Schulman Butterworth, London, 1957, pp.267-271.

105. Buckenham, M.H. and Schulman, J.H.: Molecular association in flotation. Trans. AIME, 226, 1-6 (1963).

106. Schulman, J.G.: Oppositely charged mixed collectors in flotation. Surface Chem., 9-19 (1965).

107. Pomianowski, A. and J. Leja: Equimolar solutions of xanthate and alkyl trimethyl ammonium bromide adsorption on copper, nickel and sphalerite powders. Trans. AIME, 229, 307-312 (1964).

108. Scowen, R.V. and J. Leja: Spectrophotometric studies on surfactants. Can. J. Chem., 45, 2821-34 (1967).

109. Tipman, R.N. and J. Leja: Optical measurements of adsorption in a froth flotation system. J. Colloid Interface Sci., 29, 305-313; (Schulman Memorial number) (1969).

110. Finkelstein, N.P.: Influence of alkyl trimethyl ammonium halides on the stability of potassium ethyl xanthate in aqueous solution. J. Appl. Chem., 19, 73-76 (1969).

CHAPTER 8

ANALYTICAL CHEMICAL APPLICATIONS

Xanthates have been extensively employed as reagents in chemical analysis for the separation and quantitative determination of cations of transition metals by taking advantage of the low solubilities of transition metal xanthates, and for the analysis of alcohols and carbon disulfide by their quantitative conversion to xanthate. At one time it was suggested[1,2] that alkali metal xanthates could be used as reagents for group separation in qualitative analysis, and several advantages were claimed for xanthates over hydrogen sulfide. This method, however, has not so far found favor with analytical chemists. The only mention of xanthate in qualitative analysis is in connection with the detection of molybdate ion with which it gives a red-purple color as a result of the formation of a compound of the formula $MoO_2[SC(SH)(OC_2H_5)]_2$. The test is specific for molybdates, although copper, cobalt, nickel, iron, and chromium interfere.

SEPARATION AND DETECTION OF ALCOHOLS AS XANTHATES

Paper chromatography of alcohols as the potassium xanthates was recommended by Kariyone and co-workers.[3] This procedure is based on the conversion of the alcohol to the xanthate by carbon disulfide and potassium hydroxide. A few drops of the alcohol are mixed with 0.5 ml of carbon disulfide and 0.1 g of potassium hydroxide and the mixture shaken. The liquid is decanted into another test tube and evaporated. The resultant xanthate is dissolved in 1 drop of water and the solution is developed by paper partition chromatography using alkaline butanol as the solvent. An alkaline solvent is necessary to avoid destruction of the xanthates.[4] After about 20 h the location of the xanthates on the paper is detected by their dark luminescence under ultraviolet light. In the case of solid alcohols, the xanthate formed is dissolved in formamide and developed in the same way. The chromatographic method has been found particularly suitable for the detection of traces of methanol (down to 0.1%) in ethanol. The R^F values for alcohols obtained by this method are given in Table 8-1.

TABLE 8-1

R^F Values for Alcohols Obtained by the Potassium Xanthate Method.

Alcohol	R^F Value	Alcohol	R^F Value
Methanol	0.23	Isoamyl alcohol	0.62
Ethanol	0.35	Benzyl alcohol	0.45
Isopropanol	0.44	Octyl alcohol	0.17
Isobutanol	0.54	Cyclohexanol	0.04
n-Butanol	0.55	Ethylene chlorohydrin	0.91

Spanyer and Phillips[5] adapted a chromatographic method for the separation of alcohols using a cellulose column. The alcohols are converted into potassium xanthates by standard methods. The column is first washed with 30 ml of water-saturated butanol, and 1 ml of xanthate solution containing approximately 10 mg of the solid xanthates is pipetted on the column, forced in by air pressure, and followed by solvent to fill the tube. The flow rate recommended is 6-10 drops per min.

After sufficient separation, the bands, distinguished by their fluorescence in ultraviolet light, are marked and the column is cut into sections. The individual bands are dispersed in water and filtered and washed to remove the cellulose. The filtrate is neutralized, excess thallous sulfate solution is added, and the solution boiled to coagulate the precipitate of thallous xanthate. The precipitate is filtered and dried at 80°C and the melting points are determined to identify the individual xanthates. Spanyer and Phillips have obtained good separation with 21 pairs of alcohol xanthates. The results recorded in Table 8-2 show that separation cannot be obtained only with those pairs that contain isobutyl or sec-butyl alcohols.

Roehr and Chiari[6] have described thin-layer chromatographic separation of lower alcohols by conversion into xanthates. Chromatograms are prepared on a special quality cellulose powder. Visible separation of the xanthates takes place after about 40 min, with a travel distance of 10 cm. A solvent mixture consisting of propyl alcohol, petroleum ether (boiling range 100°-150°C), and concentrated ammonia in a 3:1:1 ratio gives

TABLE 8-2

Chromatography as Potassium Xanthates of Pairs of Alcohols[a]

Excellent separation	Good separation	No separation
Methyl-ethyl	Ethyl-propyl	Butyl-isobutyl
Methyl-n-propyl	Ethyl-isopropyl	Isobutyl-sec-butyl
Propyl-isopropyl	Butyl-sec-butyl	sec-Butyl-amyl
Propyl-butyl	Butyl-amyl	
Propyl-isobutyl	Isobutyl-isoamyl	
Isopropyl-butyl	sec-Butyl-isoamyl	
Isopropyl-isobutyl		
Butyl-isoamyl		
Amyl-isoamyl		

[a] A minimum distance of 12 mm between bands is considered excellent separation, while separation of two distinct bands closer than 12 mm is called good.

the best results, the R_F values of methyl, ethyl, propyl, butyl, amyl, hexyl, and octyl alcohols being 53-90. The chromatograms are developed with a 10% nickel sulfate solution which forms a yellow or orange color. The test is sensitive to 5-10 g of alcohol.

A method for the identification of alcohols by the X-ray diffraction pattern of their xanthate derivatives has been described by Warren and Matthews.[7] This method has the advantage of ease of preparation and lack of interference from impurities formed by side reactions in the preparation of the xanthates. To obtain the X-ray pattern, copper radiation (λ = 1.542 Å) is satisfactory but chromium radiation (λ = 2.291 Å) is recommended to record lines up to 25 Å. Line intensities may be estimated visually. The index lines are given in Table 8-3 and the diffraction data are listed in Table 8-4.

A plot of the strong inner line shows an increase with chain length in homologously related structures such as those of C_4, C_5, and C_6 alcohol derivatives or of C_{10} to C_{18} compounds and C_3, C_4, and C_5 primary iso-compounds. This method of X-ray diffraction identification is applicable to the identification of a wide variety of alcohols. Patterns given as examples are for allyl, cyclohexyl, and various C_5 isomers, and the system may be extended to polyhydric alcohols.

TABLE 8-3

Index Lines of Potassium Xanthate Derivatives of Alcohols[a]

No.	Alcohol	Strongest lines, Å				Innermost line, Å
		1st	2nd	3rd	4th	
C_2	Ethyl alcohol	7.5	3.55	3.19	3.24	18.3
C_3	1-Propanol	10.1	3.00	2.21	2.65	16.4
	2-Propanol	9.6	3.06	3.19	7.6	9.6
C_4	1-Butanol	10.0	3.32	3.81	2.25	10.0
	2-Methyl-1-propanol	11.6	3.06	2.78	7.52	11.6
	2-Butanol	10.9	3.18	6.23	3.13	10.9
C_4	1-Pentanol	12.2	4.57	4.04	3.15	12.2
	3-Methyl-1-butanol	12.8	4.86	3.21	3.17	12.8
	2-Methyl-1-butanol	12.5	3.09	4.14	3.19	12.5
	3-Methyl-2-butanol	11.9	3.18	4.92	6.37	11.9
	3-Pentanol	12.1	3.17	4.62	4.07	12.1
C_6	1-Hexanol	14.0	4.59	4.07	3.18	18.0
C_7	1-Heptanol	15.6	4.62	4.06	3.14	15.6
C_8	1-Octanol	4.27	16.8	6.5	3.38	16.8
C_{10}	1-Decanol	4.29	7.3	3.67	5.14	14.6

Table 8-3 (con't)

C_{12}	1-Dodecanol	4.32	3.82	3.68	16.5	16.5
C_{14}	1-Tetradodecanol	4.33	3.81	18.2	3.68	18.2
C_{16}	1-Hexadecanol	4.35	3.70	3.83	20.5	20.5
C_{18}	1-Octadecanol	22.1	4.35	14.7	3.81	22.1
C_3	2-Propen-1-ol	9.6	3.10	2.27	4.01	9.6
C_6	Cyclohexanol	12.8	3.05	4.90	3.16	12.8
C_3	2-Methoxyeth-anol	9.4	3.04	4.32	2.88	9.4
C_4	2-Methoxymetho-xyethanol	11.4	3.16	4.44	3.26	11.4

[a] After G.G Warren and F.W. Matthews: Anal. Chem. 26, 1985 (1954); (copyright 1954 by the American Chemical Society and reproduced by permission.)

TABLE 8-4

Diffraction Data of Potassium Xanthates[a]

C_2 Ethyl alcohol		C_3 1-Propanol		C_4 1-Butanol	
d, Å	I/I_1	d, Å	I/I_1	d, Å	I/I_1
18.3	0.02	16.4	0.10	10.0	1.00
12.3	0.20	10.1	1.00	0.67	0.15
9.4	0.10	9.4	0.05	6.93	0.05
8.1	0.10	8.9	0.05	5.44	0.10
7.5	1.00	8.3	0.05	4.44	0.35
5.20	0.02	5.54	0.20	3.81	0.40
4.70	0.05	4.56	0.10	3.32	0.80
4.35	0.05	4.08	0.40	3.19	0.02
3.55	0.60	3.99	0.02	3.15	0.02
3.42	0.10	3.93	0.02	3.11	0.02
3.33	0.10	3.77	0.05	3.05	0.10
3.24	0.25	3.70	0.05	3.00	0.10
3.19	0.30	3.44	0.02	2.88	0.05
3.09	0.25	3.34	0.02	2.78	0.05
3.03	0.05	3.29	0.10	2.73	0.10
2.72	0.10	3.15	0.30	2.69	0.15
2.67	0.10	3.10	0.40	2.58	0.02

Table 8-4 (con't)

2.62	0.15	3.00	0.70	2.56	0.02
2.51	0.05	2.75	0.10	2.50	0.05
2.49	0.05	2.65	0.45	2.45	0.02
2.47	0.05	2.62	0.05	2.42	0.02
2.41.	0.20	2.57	0.02	2.40	0.15
2.36	0.10	2.50	0.10	2.32	0.05
2.33	0.10	2.39	0.30	2.30	0.05
2.27	0.15	2.28	0.30	2.25	0.40
2.24	0.05	2.21	0.50	2.18	0.05
2.19	0.05	2.03	0.15	2.16	0.10
1.63	0.05	1.99	0.15	2.12	0.10
				2.05	0.15
				1.95	0.20
				1.8	0.10

(Table 8-4 (Con't)

C_5 1-Pentanol		C_6 1-Hexanol		C_7 1-Heptanol	
d, Å	I/I_1	d, Å	I/I_1	d, Å	I/I_1
12.2	1.00	18.0	0.05	15.6	1.00
6.30	0.05	14.0	1.00	12.0	0.06
6.08	0.02	7.02	1.05	8.58	0.03
5.86	0.05	6.43	0.02	7.77	0.03
5.44	0.02	5.66	0.05	6.90	0.02
5.10	0.02	4.86	0.03	6.23	0.02
4.98	0.02	4.68	0.02	5.58	0.06
4.57	0.40	4.59	0.20	5.17	0.10
4.20	0.10	4.35	0.02	4.62	0.50
4.04	0.40	4.07	0.15	4.57	0.05
3.54	0.10	3.68	0.03	4.06	0.40
3.36	0.10	3.15	0.02	3.88	0.01
3.28	0.02	3.18	0.10	3.75	0.01
3.15	0.20	3.14	0.10	3.46	0.01
3.22	0.15	3.06	0.05	3.28	0.01
3.03	0.02	3.01	0.02	3.14	0.30
2.99	0.15	2.95	0.10	3.10	0.10
2.92	0.05	2.89	0.02	3.03	0.30
2.90	0.05	2.83	0.02	2.85	0.15
2.85	0.02	2.69	0.02	2.72	0.10
2.82	0.02	2.57	0.02	2.67	0.02

Table 8-4 (Con 't)

2.71	0.10	2.46	0.02	2.58	0.06
2.65	0.02	2.45	0.02	2.49	0.02
2.59	0.10	2.32	0.02	2.47	0.02
2.54	0.02	2.29	0.02	2.43	0.01
2.51	0.02	2.15	0.02	2.37	0.01
2.46	0.05	2.03	0.02	2.32	0.02
2.43	0.15	2.01	0.02	2.25	0.04
2.29	0.15			2.14	0.02
2.23	0.10			2.09	0.01
2.20	0.10			2.01	0.10
2.15	0.02			1.57	0.05
2.14	0.02			1.54	0.05
2.09	0.02			1.53	0.02
2.06	0.10				
2.01	0.02				
2.00	0.02				
1.98	0.05				
1.85	0.05				
1.81	0.05				
1.57	0.10				

Table 8-4 (Con't)

C_8 1-Octanol		C_{10} 1-Decanol		C_{12} 1-Dodecanol	
d, Å	I/I_1	d, Å	I/I_1	d, Å	I/I_1
16.8	0.90	14.6	0.20	16.5	0.30
12.9	0.10	9.6	0.02	11.0	0.05
9.6	0.02	7.3	0.50	8.23	0.20
8.6	0.10	6.0	0.02	6.60	0.10
7.9	0.02	5.7	0.02	5.51	0.10
6.5	0.60	5.47	0.05	5.06	0.20
5.54	0.10	5.14	0.30	4.71	0.15
5.24	0.20	4.96	0.02	4.32	1.00
5.18	0.20	4.76	0.10	4.20	0.02
4.72	0.02	4.52	0.02	4.07	0.10
4.49	0.02	4.29	1.00	3.95	0.02
4.27	1.00	4.05	0.20	3.82	0.70
4.05	0.20	3.81	0.25	3.68	0.40
4.90	0.10	3.67	0.40	3.71	0.20
3.85	0.10	3.60	0.25	3.53	0.05
3.80	0.05	3.47	0.10	3.39	0.15
3.72	0.02	3.36	0.05	3.30	0.05
3.66	0.02	3.30	0.05	3.25	0.05
3.57	0.30	3.26	0.10	3.20	0.10
3.50	0.40	3.10	0.05	3.08	0.02

Table 8-4 (Con't)

3.38	0.50	2.99	0.05	3.02	0.05
3.22	0.10	2.93	0.02	2.92	0.02
3.14	0.02	2.86	0.02	2.83	0.10
3.04	0.02	2.79	0.02		
2.98	0.02	2.73	0.02		
2.92	0.10	2.64	0.02		
2.86	0.02	2.56	0.02		
2.78	0.05	2.51	0.02		
		2.45	0.05		
		2.36	0.20		
		2.24	0.10		
		1.84	0.10		
		1.77	0.05		

Table 8-4 (Con't)

C_{14} 1-Tetradecanol		C_{16} 1-Hexadecanol		C_{18} 1-Octadecanol	
d, Å	I/I_1	d, Å	I/I_1	d, Å	I/I_1
18.2	0.60	20.5	0.40	22.1	1.00
12.2	0.05	13.7	0.20	14.7	0.80
9.1	0.15	10.2	0.05	11.0	0.10
7.3	0.05	8.2	0.05	8.8	0.10
6.1	0.15	6.8	0.15	7.33	0.50
5.22	0.02	5.80	0.05	6.27	0.15
5.00	0.20	5.27	0.02	5.51	0.15
4.67	0.10	4.96	0.20	4.95	0.30
4.33	1.00	4.67	0.20	4.67	0.15
4.20	0.02	4.35	1.00	4.35	0.90
4.13	0.02	4.11	0.10	4.16	0.20
4.01	0.20	4.02	0.10	3.98	0.20
4.81	0.80	3.83	0.50	3.81	0.70
4.68	0.50	3.70	0.70	3.70	0.50
3.61	0.20	3.64	0.15	3.53	0.10
3.54	0.05	3.58	0.05	3.44	0.02
3.44	0.30	3.50	0.15	3.34	0.02
3.32	0.15	3.39	0.05	3.20	0.02
3.25	0.02	3.36	0.05	3.02	0.05
3.19	0.02	3.27	0.10	2.92	0.02
3.12	0.05	3.18	0.02	2.82	0.05

Table 8-4 (Cont)

3.02	0.02	3.11	0.02	2.75	0.05
2.97	0.02	3.03	0.10		
2.78	0.10				

Table 8-4 (Con't)

C_3 2-Propanol		C_4 2-Methyl-1-propanol		C_4 2-Butanol	
d, Å	I/I_1	d, Å	I/I_1	d, Å	I/I_1
9.6	1.00	11.6	1.00	10.9	1.00
7.6	0.25	7.52	0.30	10.0	0.30
6.25	0.10	5.00	0.15	8.19	0.05
5.98	0.05	4.92	0.15	6.84	0.05
5.47	0.10	4.15	0.20	6.60	0.05
4.49	0.10	3.87	0.05	6.23	0.45
4.33	0.02	3.77	0.05	5.72	0.15
3.99	0.02	3.63	0.10	5.00	0.15
3.89	0.02	3.36	0.20	4.58	0.15
3.51	0.02	3.16	0.20	3.96	0.05
3.38	0.10	3.06	0.60	3.80	0.05
3.33	0.10	2.91	0.20	3.59	0.10
3.29	0.05	2.78	0.40	3.50	0.10
3.19	0.30	2.51	0.05	3.45	0.05
3.17	0.10	2.48	0.05	3.40	0.05
3.06	0.50	2.46	0.10	3.32	0.05
3.02	0.10	2.40	0.05	3.24	0.63
2.94	0.02	2.35	0.02	3.18	0.50
2.89	0.10	2.33	0.05	3.13	0.45
2.83	0.02	2.31	0.05	3.10	0.30

Table 8-4 (Con't)

2.79	0.20	2.26	0.30	2.87	0.10
2.72	0.05	2.15	0.10	2.76	0.05
2.60	0.02	2.08	0.10	2.70	0.05
2.49	0.02	2.04	0.05	2.55	0.02
2.44	0.02	1.94	0.05	2.47	0.10
2.41	0.30	1.87	0.10	2.39	0.05
2.34	0.02	1.57	0.02	2.35	0.05
2.31	0.02	1.53	0.05	2.24	0.05
2.29	0.02	1.52	0.02	2.21	0.05
2.25	0.02			2.17	0.05
2.23	0.10			2.11	0.20
2.20	0.02			1.94	0.05
2.15	0.02				
2.06	0.15				
2.05	0.20				
1.99	0.02				
1.97	0.02				
1.95	0.02				
1.90	0.05				
1.70	0.02				

Table 8-4 (Con't)

C_5 3-Methyl-1-butanol		C_5 2-Methyl-1-butanol		C_5 3-Methyl-2-butanol	
d, Å	I/I_1	d, Å	I/I_1	d, Å	I/I_1
12.8	1.00	12.5	1.00	11.9	1.00
9.4	0.02	7.8	0.10	10.8	0.02
7.6	0.02	6.20	0.05	6.90	0.02
6.4	0.05	5.32	0.05	6.37	0.20
5.32	0.05	5.09	0.15	6.20	0.10
4.86	0.50	4.63	0.02	5.82	0.02
4.29	0.10	4.14	0.40	5.40	0.02
4.23	0.10	3.93	0.02	5.15	0.02
4.13	0.15	3.67	0.10	4.92	0.25
3.79	0.05	3.44	0.05	4.83	0.02
3.52	0.05	3.19	0.30	4.20	0.10
3.47	0.05	3.09	0.60	3.97	0.15
3.34	0.02	2.93	0.05	3.78	0.10
3.21	0.30	2.84	0.20	3.60	0.02
3.17	0.30	2.54	0.15	3.38	0.02
3.02	0.10	2.46	0.05	3.27	0.15
3.00	0.10	2.33	0.20	3.18	0.15
2.98	0.10	2.23	0.02	3.10	0.02
2.81	0.02	2.15	0.10	3.05	0.02
2.72	0.10	2.07	0.05	2.98	0.02
2.69	0.05	2.04	0.05	2.95	0.15
2.56	0.20	1.98	0.02	2.92	0.02

Table 8-4 (Con't)

2.49	0.05	1.59	0.02	2.80	0.10
2.46	0.05	1.55	0.05	2.52	0.05
2.42	0.05			2.46	0.05
2.37	0.05			2.33	0.05
2.33	0.05			2.26	0.05
2.27	0.15			2.15	0.05
2.19	0.05			2.11	0.05
2.14	0.15			2.09	0.05
				1.99	0.05
				1.60	0.06
					0.05

Table 8-4 (Con't)

C_5 3-Pentanol		C_3 2-Propen-1-ol		C_6 Cyclohexanol	
d, Å	I/I_1	d, Å	I/I_1	d, Å	I/I_1
12.1	1.00	9.6	1.00	12.8	1.00
6.3	0.10	9.1	0.10	9.2	0.05
5.90	0.10	5.44	0.25	7.3	0.02
5.43	0.02	4.83	0.05	6.8	0.02
4.93	0.02	4.56	0.05	6.1	0.05
4.62	0.30	4.24	0.05	5.71	0.05
4.22	0.05	4.01	0.30	5.25	0.02
4.07	0.30	3.82	0.05	4.90	0.10
3.60	0.08	3.67	0.10	4.27	0.05
3.41	0.08	3.60	0.10	4.23	0.05
3.17	0.40	3.37	0.02	3.91	0.05
3.00	0.20	3.21	0.08	3.76	0.02
2.92	0.20	3.16	0.20	3.43	0.02
2.87	0.02	3.07	0.08	3.16	0.10
2.72	0.10	3.01	0.80	3.05	0.15
2.66	0.10	2.85	0.05	2.98	0.05
2.56	0.01	2.64	0.25	2.94	0.05
2.48	0.02	2.57	0.05	2.78	0.02
2.45	0.02	2.48	0.25	2.69	0.05
2.41	0.05	2.40	0.05	2.60	0.02
2.33	0.10	2.27	0.40	2.51	0.02
2.21	0.08	2.21	0.02	2.42	0.02

Table 8-4 (Con't)

2.04	0.08	2.16	0.05	2.17	0.02
2.01	0.05	2.01	0.20		
		1.96	0.10		
		1.91	0.10		
		1.57	0.10		
		1.55	0.20		
		1.50	0.15		

Table 8-4 (Con't)

C_5 2-Methoxyethanol		C_4 2-Ethoxyethanol		C_4 2-Methoxymethoxyethanol	
d, Å	I/I_1	d, Å	I/I_1	d, Å	I/I_1
9.4	1.00	10.7	1.00	11.4	1.60
7.5	0.05	7.5	0.10	7.73	0.05
5.51	0.15	5.59	0.20	6.49	0.15
4.51	0.30	4.76	0.10	5.70	0.20
4.43	0.40	4.42	0.10	5.49	0.05
3.86	0.10	4.16	0.15	5.05	0.30
3.50	0.10	3.91	0.05	4.89	0.05
3.31	0.10	3.66	0.30	4.44	0.40
3.27	0.08	3.58	0.05	4.13	0.10
3.26	0.08	3.37	0.05	3.86	0.10
3.21	0.20	3.28	0.25	3.70	0.15
3.17	0.10	3.12	0.60	3.61	0.10
3.12	0.05	2.79	0.25	3.42	0.05
3.04	0.30	2.61	0.15	3.26	0.35
2.98	0.30	2.54	0.10	3.16	0.50
2.92	0.05	2.48	0.10	3.11	0.30
2.88	0.35	2.41	0.10	2.87	0.20
2.82	0.05	2.35	0.20	2.83	0.05
2.68	0.12	2.30	0.05	2.80	0.05
2.65	0.10	2.21	0.08	2.77	0.05

Table 8-4 (Con't)

2.60	0.15	2.11	0.15	2.56	0.05
2.55	0.10	2.06	0.15	2.46	0.05
2.53	0.10			2.39	0.20
				2.26	0.10
				2.22	0.05
				2.19	0.15
				2.16	0.15
				2.07	0.15

[a] After G.G. Warren and F.W. Matthews: Anal-Chem 26, 1985 (1954) (Copyright 1954 by the American Chemical Society and reproduced by permission.)

DETERMINATION OF CARBON DISULFIDE BY CONVERSION TO XANTHATES

The oxidation of xanthate by chloramine-T (see Chapter 3) has been adapted for an elegant method for determining carbon disulfide.[8] A weighed amount of a liquid containing carbon disulfide is mixed with 5-6 ml of freshly prepared alcoholic potash (10%) in a stoppered flask. The xanthate formed is dissolved in water and mixed with a known excess of acidified chloramine-T solution. To the mixture about 25 ml ethyl alcohol are added. The reactants are set aside for at least 1/2 h. At the end of this period 10 ml of 10% potassium iodide solution are added, and the liberated iodine titrated against standard thiosulfate solution. According to the oxidation reaction of xanthate by chloramine-T, 14 equivalents of the oxidant are consumed per mole of carbon disulfide.

$$CS_2 \xrightarrow[EtOH]{KOH} S{=}C\begin{matrix}\diagup SK \\ \diagdown OEt\end{matrix} \xrightarrow[3H_2O]{7O} EtOH + H_2SO_4 + KHSO_4 + HCOOH$$

The accuracy of this method has not been reported. The solubility of carbon disulfide determined in this way, however, reasonably agrees well with the value reported in the literature, and the method can be considered satisfactory for routine estimation of carbon disulfide in mixtures that do not contain reducing impurities.

A complexometric determination of carbon disulfide by converting it into xanthate has been described by Kakedy and Makkay.[9] A gaseous mixture containing carbon disulfide is bubbled through aqueous 25% potassium hydroxide solution, anhydrous calcium chloride, and then through two wash bottles containing 50 ml of 20% potassium hydroxide solution in ethanol. The solutions in the two wash bottles are combined, and 15 ml of 10% acetic acid and 15-40 ml of 0.005 M nickel chloride solution at pH 3 are added. The precipitate of nickel xanthate formed is filtered and the excess nickel determined by titration with 0.01-0.005 M EDTA at pH 9-10 in the presence of murexide indicator. The method is very rapid and the error is only ± 0.6%.

COLORIMETRIC DETERMINATION OF COBALT AS ETHYL XANTHATE

This procedure[10] is based on the extractability of ethyl xanthates from the aqueous phase by organic solvents such as benzene, carbon tetrachloride, chloroform, ethylene chloride, and ligroin. Among the ethyl xanthates of transition metal ions those of copper, silver, mercury, cadmium, and uranyl are not extracted by organic solvents, while those of zinc, iron, nickel, vanadium, tin, antimony, bismuth, and lead remain in the aqueous phase if it is rendered strongly alkaline. From the alkaline aqueous phase only cobalt ethyl xanthate is extracted by organic solvents. This affords a method for the separation and determination of cobalt in the presence of other transition metal ions. The final analysis is done by a colorimeter. In a mixture containing cobalt and nickel, the precipitation with ethyl xanthate is carried out in a slightly acidified ($pH \simeq 6$) aqueous solution. The precipitate of mixed xanthates is extracted with carbon tetrachloride, and the organic liquid layer is separated and treated with ammonium hydroxide. The nickel xanthate goes into the aqueous phase. Cobalt ethyl xanthate in the carbon tetrachloride layer is estimated by colorimeter. If ferric or ferrous ions are present in large quantities it is advisable to precipitate them with ammonium tartrate.

8. ANALYTICAL CHEMICAL APPLICATIONS

This method is applicable to the chemical analysis of ores and alloys in which cobalt is one of the constituents. As may be expected, the accuracy is not better than ±2%, but is probably satisfactory for routine industrial analysis.

Kakovskii[11] has explored the possibilities of selective precipitation of cobalt with xanthate in aqueous media. In the presence of other metallic ions, cobalt is precipitated almost simultaneously with nickel, before zinc and after copper and cadmium. This indicates a possibility of selective separation of cobalt from metal ions other than nickel. The separation from nickel could then be achieved by the extraction of nickel xanthate in carbon tetrachloride from an ammoniacal medium as described above.

An alternate method of separating cobalt from nickel has been described by Oyama and coworkers.[12,13] A mixed precipitate of cobalt and nickel is washed with a concentrated solution of ammonia. Cobalt xanthate remains as an unaltered precipitate, and nickel xanthate dissolves in the ammonia solution, giving a nickel ammine complex. It has been possible to employ xanthate as a selective precipitant for nickel and cobalt ions. The separation of cobalt and nickel xanthates from the other metallic xanthates is accomplished by taking advantage of their strong hydrophobic character, which allows the separation to be made by flotation. The precipitation of cobalt xanthate is to some extent hindered by the coexistence of zinc and ferrous ions, but agitation ensures up to 90% recovery of cobalt when potassium methyl xanthate is used. After the separation of cobalt xanthate from nickel xanthate by ammonia washing, nickel can be recovered from the ammine solution by precipitation as nickel sulfide. The filtrate remaining after the nickel sulfide is removed contains ammonia and xanthate. The xanthate can be recovered by carefully concentrating the filtrate. Alternatively the mixed precipitate of cobalt and nickel xanthates may be washed with sodium hydroxide or calcium hydroxide (in place of the ammonia solution), which converts nickel xanthate into nickel hydroxide, and the xanthate of sodium or potassium can be recovered in the solution. This procedure has been employed for the separation of cobalt and nickel from the leaching solution of pyrite cinder decopperized by scrap ion.

EXTRACTION OF COBALT, ZINC, AND CADMIUM IONS

The separation of these ions can be accomplished[14] by selective extraction of their xanthates in chloroform. A 3:1 volume ratio of aqueous to chloroform phase is used. The xanthate is used in 100- to 500-fold excess. With potassium xanthate at pH 1, only cobalt is extracted, and at pH 10 cadmium is separated from much of the associated zinc. By using benzyl xanthate in place of ethyl xanthate practically all the cobalt is extracted at pH 0.85-5.2 and zinc and cadmium are extracted in the pH range 1.9-5.2. Cobalt can thus be separated in the pH range 0.85-1.9.

GRAVIMETRIC DETERMINATION OF NICKEL

The precipitation of nickel ethyl xanthate is quantitatively carried out at pH 3-4.[15] The precipitate is easily filtered and can be dried at temperature below 150 $^{\circ}$C. In the gravimetric determination of nickel, the solution is acidified with hydrochloric acid until the pH is lowered down to 3-4 as tested by methyl orange indicator. Potassium ethyl xanthate solution (0.2%) is added in small portions until a small excess is present. The contents are agitated slowly for 15 min and then filtered. The precipitate is washed with dilute ethyl xanthate solution and then with a little water. It is dried for 30 min at 105°C and weighed. The presence of other transition metal ions would obviously interfere, and therefore the applicability of this method is restricted to the analysis of solutions that may contain only alkali and alkaline earth metal ions. Satisfactory results (±1% accuracy) are obtained for solutions containing 1-45 mg of nickel.

COLORIMETRIC DETERMINATION OF MOLYBDENUM

The red-purple-colored complex formed by the combination of ethyl xanthate and hexavalent molybdenum has been used for the colorimetric determination of molybdenum.[16] The pH is adjusted in the range 1.11-1.56 to give a maximum and definite optical density. The color develops after shaking and remains stable for about 1 min only. Stability can be improved by extraction with petroleum ether or toluene. Beer's law is followed up to a concentration of 0.4 mg per liter.

SEPARATION OF TECHNETIUM FROM RHENIUM

Kiba and co-workers[17] have described a liquid-liquid extraction with potassium xanthate for the separation of technetium from rehnium. The method is based on the formation of a compound by treating pertechnetate ion with potassium xanthate. The nature of this compound has not been analyzed, but it is known to be soluble in carbon tetrachloride or chloroform. A technetium solution (2.0 mg, containing rhenium) is mixed with 1.5 N hydrochloric acid (2.0 ml) and 1 ml of potassium xanthate solution containing a minimum of 60 mg of xanthate per 20 g of technetium. After mixing in a separatory funnel, 5 ml of carbon tetrachloride are added and the mixture shaken for 20 min. The organic layer is transferred to a stainless steel dish and evaporated to dryness under an infrared lamp. Rhenium remains entirely in the aqueous phase, and thus separation from technetium is achieved. This is probably attributable to the fact that rhenium is not reduced by xanthic acid to the same oxidation state as technetium because of the difference in redox potentials between perrhenate and pertechnetate. The separation of technetium is affected by the presence of copper(II) and nickel(II) ions but only slightly by vanadium(V), chromium(III), cobalt(II), arsenic(III), molybdenum(VI), antimony(III), tungsten(VI), and platinum(IV), each at a concentration of 0.02 mg per liter. The method has potential applications in radioactivation analysis.

AMPEROMETRIC DETERMINATION OF LEAD IN THE PRESENCE OF MERCURY OR SILVER

This determination, described by Deshmukh and Saraswathi,[18,19] is based on the fact that the half-wave potential of lead is considerably more negative than that of mercury or silver and, hence, the latter may be determined cathodically while lead is determined anodically. Aliquots of mercury and lead solutions are mixed with 0.05 M ammoniacal tartrate buffer (pH $\simeq$ 9) used as the supporting electrolyte. The current is noted at -0.2 V and the titration with potassium ethyl xanthate is carried out at this potential. The end points for mercury and lead are located at the intersections of the line of the residual current with the reaction line and with the excess reagent line, respectively. An almost identical procedure is employed for the analysis of lead ions in a solution containing silver in place of mercury. The method is applicable to milligram quantities of cations, and the accuracy is within 1%.

REFERENCES

1. Wenger, P., R. Duckert, and E.Ankadji: New methods of analysis for cations based on the use of potassium ethyl xanthate. Helv. Chim. Acta, 28, 1992-1609 (1945); Chem. Abstr., 40, 1410, (1946).

2. Lavin, L.R.C.: A general method of separation and identification of cations by using potassium ethyl xanthate. Inform. Quim. Anal. (Madrid), 5, 62-67, 98-103 (1951); Chem. Abstr., 46, 5486h, (1952).

3. Kariyone, T., Y. Hashimoto, and M. Kimura: Paper partition chromatography of alcohols with the potassium xanthogenates. Nature, 168, 511 (1951).

4. Gasparic, J. and J. Borecky: Paper chromatographic separation of alcohols as their xanthates. J. Chromatog., 4, 138-143 (1960).

5. Spanyer, J.W. and J.P. Phillips: Chromatographic separation of alcohols as xanthates. Anal. Chem., 28, 253 (1956).

6. Roehr, M. and D. Chiari: Thin layer chromatographic separation of xanthates of lower alcohols. Mikrochim. Acta, 137-139 (1967).

7. Warren, G.G. and F.W. Matthews: X-ray diffraction identification of alcohols by their xanthate derivatives. Anal. Chem., 26, 1985-1987 (1954).

8. Rao, V.R.S. and A.R.V. Murthy: Determination of carbon disulfide by chloramine-T. Talanta, 4, 206-209 (1960).

9. Kekedy, L. and F. Makkay: Complexometric determination of xanthates and of carbon disulphide. Studia Univ. Babes-Bolyai, Ser. 1, 139-144 (1962); Chem. Abstr., 61, 12, 641 (1964).

10. Pilipenko, A.T. and N.V. Ul'ko: Colorimetric determination of cobalt as ethyl xanthate. Zh. Anal. Khim., 10, 299-304 (1955).

11. Kakovskii, A.I.: Theoretical foundations of xanthate method of precipitating cobalt from solutions. Tsvetn. Metal., 30(7), 42-51 (1957); 32(8), 57-61 (1959).

12. Oyama, T., J. Shimoiizaka, T. Yamasaki, and I. Baba: Xanthate as a selective precipitant for nickel and cobalt ions. Nippon Kogyo Kaishi 78, 391-396 (1962).

13. Yamasaki, T., J. Shimoiizaka, H. Sasaki and T. Oyama: Separation of nickel from cobalt. Nippon Kogyo Kaishi, 79(2), 25-32 (1962).

14. Sheyanova, P.R., A.G. Airapetyan, S.A. Ryabova and V.M. Ryabov: Extraction of cobalt, sinc and cadmium xanthates. Tr. Khim. Khim. Tekhnoh. No. 2,410-413 (1959).

15. Kekedy, L. and F. Makkay: Gravimetric determination of nickel. Studia Univ. Babes-Bolyai, Ser. 1, 135-138 (1962). Chem. Abstr., 61, 2471, (1964).

16. Shu-Chuan Liang and Pao-Yun Hsu: Colorimetic determination of molybdenum with ethyl xanthate. Hua Hsueh Hsueh Pao, 22, 171-176 (1956).

17. Kiba, T., N. Terada, N. Okawa, and S. Osaki: Separation of technetium from rhenium by liquid-liquid extraction with potassium xanthate. Talanta, 13, 1385-1388 (1966).

18. Deshmukh, G.S. and Saraswathi: Amperometric determination of mercury and lead with xanthate and their separation. Indian J. Chem., 3, 489-490 (1965).

19. Deshmukh, G.S. and K. Saraswathi: Amperometric determination of silver as xanthate and its determination in presence of lead. Bull. Chem. Soc. Japan, 40, 1545-1546 (1967).

CHAPTER 9

XANTHATES IN THE CELLULOSE INDUSTRY

The introduction of xanthates in the manufacture of rayon, starting from cellulose derived from wood pulp or cotton linters, originated with the discovery of cellulose xanthate by Cross and Bevan.[1] Cellulose treated with alkali, by a process known as mercerization in the textile industry, reacts with carbon disulfide to form cellulose xanthate:

$$\underset{\text{Alkali cellulose}}{(\text{Cellulose})\text{ONa}} + CS_2 \longrightarrow \underset{\text{Cellulose xanthate}}{SC\begin{matrix} \diagup \text{O(Cellulose)} \\ \diagdown \text{S-Na} \end{matrix}} \qquad (9\text{-}1)$$

The exact chemical composition of cellulose depends upon the source from which it is derived; the simplest formula is $C_6H_{10}O_5.OH$.

Cellulose xanthate is soluble in dilute sodium hydroxide, producing a highly viscous orange-colored solution called viscose. From this solution the regenerated cellulose is precipitated by the action of an acid.

9. XANTHATES IN THE CELLULOSE INDUSTRY

The industrial process is carried out as follows. Wood pulp is treated with a 17-18% solution of sodium hydroxide in a steeping press and allowed to soak for about 1 h. The object of this procedure is to separate α cellulose, which remains insoluble, while β and γ -cellulose are leached out by the alkali. The sheets of cellulose are then pressed to squeeze out the alkali. The resultant alkali cellulose is then allowed to age for a period of several days in a steel compartment having provision for maximum air contact. While ageing, the alkali cellulose is oxidized and the length of the polymeric chain is reduced. This ageing step governs the viscosity of the final viscose solution. The properly aged alkali cellulose is treated with carbon disulfide for several hours in a churn which rotates around its horizontal axis. As the carbon disulfide is absorbed, the color of the cellulose changes from white through gradations of yellow to orange. The orange color indicates the completion of the xanthating process. The excess carbon disulfide vapors are drawn off and the xanthate is discharged into a viscose mixer in which it is dissolved in a dilute solution of sodium hydroxide.

The resulting product contains about 7.25% cellulose and 6.5% sodium hydroxide. Because of its highly viscous character it is named viscose. It is then filtered to remove any foreign matter and undissolved cellulose and stored in tanks for several days at a temperature controlled at 16°C. This process is known as viscose ripening. To test the degree of ripeness of the viscose, 1 drop is placed in a salt solution and its coagulation observed. The salt solution is made up of x ml of a concentrated stock solution of sodium chloride and (10 - x) ml of water; the resulting 10 ml are diluted with water to make a total of 40 ml. The value of x is changed in successive tests until the concentration of salt solution that just causes the viscose to coagulate is found. This value of x is the salt index. A drop of viscose in the correct state of ripening divides into three flocks. A viscose that requires more salt than customary for proper coagulation is not ripe enough and should be held in the ripening tank for a further period of time. The salt index is generally about 4.0. Variation in ripeness of the viscose causes nonuniformity in a number of physical properties such as tensile strength, stress-chain characteristics, spinning properties, and dye acceptance of the resulting yarn.

The ripened cellulose xanthate solution is converted to a textile fiber in the spinning operation. The viscose solution is pumped to a spinning machine and delivered to the spinneret, which consists of a small nozzle with many minute holes immersed in a coagulating bath into which the fine streams of viscose solution are extruded. A typical spinning bath is composed of warm water at 40°-45°C containing 8-10% sulfuric acid, $13\frac{1}{2}$-20% sodium sulfate, approximately 1% zinc sulfate, and 4-10% glucose. It may also contain a small amount of a cationic surface-active agent to lessen clogging of the spinneret openings. As the viscose enters this bath in fine streams it coagulates instantly. The principal action of sulfuric acid is to regenerate the cellulose:

$$2(\text{Cellulose})\,\text{O}\overset{\overset{\displaystyle S}{\|}}{\text{C}}\text{—SNa} + \text{H}_2\text{SO}_4 \longrightarrow 2(\text{Cellulose})\cdot\text{OH} + \text{Na}_2\text{SO}_4 + 2\text{CS}_2 \qquad (9\text{-}2)$$

EMULSION XANTHATION OF CELLULOSE

In recent years a technique has been developed for the xanthation of cellulose by emulsified carbon disulfide. In this process the cellulose derivative is completely solubilized by 100 parts 12-18% sodium hydroxide. Samples of pulp are torn by hand into small pieces (about 5 mm X 5 mm), and conditioned for 2 h at 20°C and 65% relative humidity. Samples in small batches are treated with desired quantities of sodium hydroxide solutions of various concentrations with agitation to disintegrate the pulp, after which carbon disulfide (100 ml per 50 g of sample) is added. The contents are agitated for 6 h at 20° ± 0.1°C, after which they are left to stand for 16-17 h to allow any undissolved residue to settle.

This technique was first studied by Samuelson and co-workers.[2-5] It was shown that the amount of insoluble residue decreases to nearly zero when the concentration of sodium hydroxide is increased to 100 g/litre in the case of normal and hot purified pulps; linters show little variation and cold purified pulps have intermediate values. Solution with carbon disulfide is very rapid at first, but becomes negligible after 6 h. Mercerization of the pulp followed by dilution and washing augments the reactivity.

The emulsion procedure results in a remarkably high degree of xanthation even at low alkali concentrations, the insoluble residue showing a value of over 0.50 and the soluble still higher. The cold alkali-treated pulp acquires a degree of xanthation higher than that of hot alkali-extracted pulp and yet yields a larger amount of insoluble residue. It is, however, more soluble than emulsion-xanthated cotton linters. For the cold alkali-treated pulp, the amount of undissolved residue after emulsion xanthation remains unchanged for alkali concentrations below 12%, and no change in the degree of xanthation is found. When the alkali concentration exceeds 12% the amount of undissolved residue decreases, the lowest value occurring at 16% alkali. The degree of xanthation increases as the concentration is increased to 20%.

The washing conditions following mercerization have a considerable effect on dissolution in emulsion xanthation. When the cellulose is washed with sodium chloride solution directly after mercerization with 25% sodium hydroxide at 20°C, dissolution is better at any sodium hydroxide concentration and no significant difference in degree of xanthation can be observed. In the case of wood pulp, when the cellulose is dried after mercerization or washed with water and then emulsion-xanthated directly without drying, the dissolution decreases. Drying at elevated temperature is not desirable because of its detrimental effect on dissolution. The degree of xanthation of the residue increases on repeated treatment with sodium hydroxide and hydrogen sulfide, and there is a marked difference between the behavior of cotton linter and wood pulp.

Some studies have been made on the xanthation of cellulose mercerized with potassium or lithium hydroxide in place of sodium hydroxide. Bartunek[6,7] observed that the resulting xanthates are not completely soluble in the respective alkalis although they are soluble in aqueous sodium hydroxide. This is attributed to the lower degree of swelling resulting from the more weakly hydrated potassium or lithium ion. Evidently the nature of the base is more significant in viscose formation than is the xanthation itself.

9. XANTHATES IN THE CELLULOSE INDUSTRY

In recent years Schmiedeknecht and Claus[8] have found that the introduction of surface-active agents results in higher rates of xanthation by emulsion technique. The particular wetting agent employed in emulsion xanthation has the trade name Sultafon-T, which is the sodium salt of a highly sulfated fatty acid and an alkane sulfonic acid. This agent tends to equalize penetration of alkali in the steeping of nonuniformly dried cellulose. The degree of swelling of the cellulose is not affected by the surfactant. Its principal function is to reduce the surface tension of the liquids, thus promoting penetration of alkali.

Alkali-free sodium cellulose xanthate powders have been recently described.[9] These are given the trade name Sup-R-Stren, abbreviated SRS. Being free of caustic, they hydrolyze to cellulose during drying to provide high tenacity and microscopically thin coatings. Nonwoven materials become permanently bonded nonblocking fabrics when printed or saturated with SRS solutions followed by drying. The preparation of such decausticized xanthate powders may be carried out following a method described by Bridgeford.[10] Commercial viscose, containing 8% cellulose, 6.6% total alkali, and

1.10% xanthate sulfur is diluted to a 4% cellulose contents, and 600 ml are placed in a regenerated cellulose dialysis bag. The bag is placed for 1 h at 15°C until completion of dialysis. The dialyzed viscose is diluted to 2% cellulose content and the pH adjusted to 11. The viscose thus decausticized is now fed through a spray dryer and treated with high-velocity heated air at 130°C to give a dry powder with a 12.5% xanthate sulfur content. Xanthate solutions can also be decausticized by cation or anion exchange, ion retardation, liquid ion exchange, or a combination of these methods.

Cellulose xanthate has been used as the starting material for the preparation of cellulose thiourethan in a method described by Allewelt and Watt.[11] Viscose is added to a solution of a zinc salt; the zinc cellulose xanthate formed as a pale yellow precipitate is filtered off, washed, and added to an aqueous solution of primary amine. The reaction between zinc xanthate and amine is followed to proceed for 2-3 days. The reaction product is a cellulose derivative containing nitrogen and sulfur.

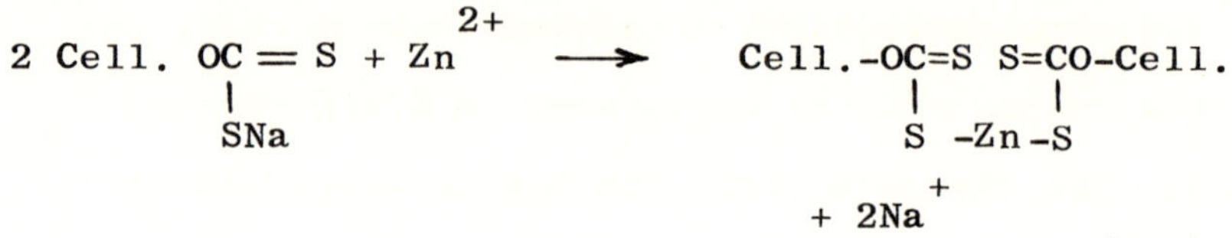

(9-3)

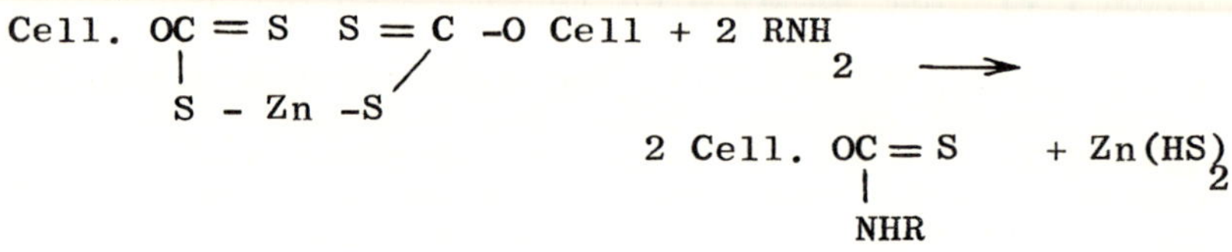

(9-4)

It is postulated that cellulose xanthic acid is formed by hydrolysis of zinc cellulose xanthate and that the free acid reacts with amine. Part of the xanthic acid decomposes into carbon disulfide.

The degree of substitution of cellulose thiorethan increases with the percentage of carbon disulfide up to 85%. An increase in the concentration of amine also promotes a higher degree of substitution. It is possible to prepare derivatives having a degree of substitution as high as 0.33.

The reaction between zinc cellulose xanthate and aniline yields a particularly useful fibrous product which gives a transparent viscous solution in dilute alkali (8% cellulose thiourethan in 3% aqueous sodium hydroxide). This material is said to have several useful properties; the more notable ones are its resistance to hydrochloric acid, low water-swelling characteristics, low water regain, and a density lower than regenerated cellulose fibers.

A copper xanthate fiber has been described recently.[12] It is prepared from viscose in two stages. In the first stage the formed filament is coagulated in a bath containing 150-250 g/liter of sodium sulfate with different amounts of sodium bicarbonate or ammonium sulfate and then separately washed in a solution of sodium sulfate to remove absorbed sulfur compounds. In the second step the fiber is brought into contact with a 30-50 g/liter of cupric sulfate solution, treated with acid, washed, and dried. The fiber now is composed of cuprous xanthate and cellulose dixanthate. This filament has a lower tensile strength than viscose, but is superior in elongation and folding resistance and exhibits bacteriostatic and bactericidal properties.

HIGHLY XANTHATED CELLULOSE XANTHATES

The xanthation of cellulose can be extended to esterify more than one alcoholic group. Geiger and Weis[13] have described the preparation of cellulose trixanthate and polyxanthates starting from fibrous cellulose. The alcoholic groups of cellulose are converted into sodium hydroxide, and the product is further treated with carbon disulfide. The reactions are represented by the equations:

$$\left[C_6H_7O_2\text{-}(OH)_y\ (OCS_2Na)_x\right]_n + nyNaOH \rightleftharpoons \left[C_6H_7O_2(ONa)_y(OCS_2Na)_x\right]_n + nyH_2O \qquad (9\text{-}5)$$

$$\left[C_6H_7O_2(ONa)_y\ (OCS\ Na)\right]_n + nyCS \rightleftharpoons \left[C_6H_7O_2(OCS_2Na)_{x+y}\right]_n \qquad (9\text{-}6)$$

In these equations, x + y = 3 when the alkali concentration is increased, and the equilibrium is moved to the right; at equal carbon disulfide concentration the degree of xanthation increases and in order to shift the second equilibrium to the right the concentration of carbon disulfide must be increased considerabley; generally its molar quantity should be twice that of the cellulose.

The percentage of esterification of cellulose by xanthate is termed the γ number. Its value depends upon various factors such as the source of cellulose[14] and the temperature of the xanthation reaction.[15,16] With 100% carbon disulfide (based on cellulose), the value of γ for cotton cellulose is 50-58, that of rayon cellulose 65, and that of xylan prepared from straw only 40. The γ value can be increased by the addition of more carbon disulfide.

The dependence of the γ number as a function of the time of xanthation passes through a maximum owing to the simultaneous esterification of alkali cellulose and the decomposition of cellulose xanthate. An increase in temperature increases the velocity of both reactions. In the temperature interval 20°-30°C, an increase in temperature with a simultaneous decrease in the time required for xanthation does not affect the degree of esterification. The γ number varies from 55 at 20° to 50 at 30°. The distribution of carbon disulfide at different temperatures in the range 20°-30°C at a constant γ number is the same. In the range 0°-40°C, the dependence of a maximum γ number on the reaction temperature is more pronounced; γ = 70 at 10° and 48 at 40° with analogous variations in the distribution of carbon disulfide.

TECHNIQUES FOR THE ANALYSIS OF CELLULOSE XANTHATES

In the course of xanthation of cellulose, knowledge of the degree of substitution (DS) and the degree of polymerization (DP) are often required. Various methods have been described to determine these values quantitatively.[17-19]

The degree of polymerization is determined by measuring the viscosity of the solution of cellulose xanthate in ethylene chlorohydrin or dimethyl sulfoxide. Two relations between intrinsic viscosity $[\eta]$, DP value, and the γ number were given by Schurz and Zimmerl:[20]

$$[\eta]\ (\text{in } HOCH_2CH_2Cl) = (12.7 - 0.0783\ \gamma)10^{-3}\ DP$$

$$\text{and } [\eta]\ (\text{in } (CH_3)_2SO = (12.2 - 0.0783\ \gamma)10^{-3}\ DP$$

A more rigorous analysis has been made by Elmgren.[21] In his procedure a film of regenerated cellulose is dissolved in cadoxen. The solution is diluted with an equal volume (10 ml) of water and filtered, and its intrinsic viscosity is determined. The intrinsic viscosity is related to the weight average degree of polymerization ($\overline{DP}_w$) by the expression

$$[\eta]\ \text{Cden} = 0.0184\ \overline{DP}_w 0.76$$

which is based on a viscometric analysis performed earlier.

Elmgren has also shown a relationship between the intrinsic viscosity of homogeneously xanthated sodium cellulose xanthate, $[\eta]$ Cden, and the degrees of substitution in primary ($DS_{2,3}$) and in secondary (DS_6) positions. This relationship is given by

$$[\eta] = 1.25\ [\eta]\ \text{Cden}^{1.08} + 0.88 - DS_{2,3} - 0.29(DS_6)\ [\eta]\ \text{Cden}^{1.24}$$

This relation illustrates quantitatively the negative dependence of DS_6 on $[\eta]$ and also provides a method for determining the Dsi distribution from combined measurements of $[\eta]$, $[\eta]$ Cden, and DS (total degree of substitution; which may be determined by the method to be described below. Rearrangement of the above expression results in the relation

$$DS_{2,3} = 0.86[\eta]\ /\ [\eta]\ \text{Cden} + [\eta]\ \text{Cden}^{-0.24}$$

$$-1.07[\eta]\ \text{Cden}^{-0.16} + 0.25DS$$

As the viscosity dependence on $DS_{2,3}$ is completely different from its dependence on DS_6, this method for determining the distribution of $DS_{2,3}$ is very accurate.

9. XANTHATES IN THE CELLULOSE INDUSTRY

Earlier methods for determining the percentage of xanthate sulfur or the γ- value of cellulose xanthates largely ignored interference from other sulfur compounds. The analysis of xanthates after separation of the by-product compounds was first described by Samuelson and Gärtner.[22] In their procedure the by-product compounds are removed by passing a dilute solution of viscose through an anion exchanger, which adsorbs the by-product ions. The cellulose xanthate ions are easily washed through the ion-exchange column and the xanthate concentration determined by iodimetric titration (see p. 97). This method has been modified by Dux and Phifer,[23] who have employed spectrophotometric methods for the determination of xanthate separated from the by-products, (see p. 110). Viscose is passed through a specially designed ion-exchange column. First, 3-4 ml are discarded to avoid dilution by the water present in the resin. One to 3 drops of the viscose are then collected in a weighed 50-ml volumetric flask containing about 20 ml of 0.1 N sodium hydroxide. The flask is weighed again and then shaken thoroughly to dissolve the viscose. The solution is made up to volume with

0.1 N sodium hydroxide and chilled in an ice bath. Five milliliters of the solution are then diluted to 250 ml with 0.1 N sodium hydroxide, and the optical density of this solution measured using 0.1 N sodium hydroxide as a reference. Not more than 1 h should elapse between the sample collection and reading, even if the sample is maintained at 0°C. To calculate the percentage of xanthate sulfur, the molar extinction coefficient of xanthate has been determined by using viscose of known xanthate concentration. A value of 15,900 has been obtained by Dux and Phifer. By using this value the following formulas may be used for calculating the percentage of xanthate sulfur or γ - value of the viscose when determined according to the procedure described above.

$$\%\ \text{Xanthate sulfur} = \frac{2.5 \times 64 \times 100 \times \text{Optical density}}{15{,}900 \times \text{Sample weight}}$$

$$\gamma\ \text{Value} = \frac{\text{Optical density} \times 50 \times 162 \times 100}{15{,}900 \times 20 \times \text{Sample weight} \times \%\ \text{cellulose in viscose}}$$

where 162 is the molecular weight of the glucose monomer unit of cellulose. The percentage of xanthate sulfur is calculated on the weight of a sample of ion-exchanged viscose and not on the original viscose. However, the error thus introduced is less than 2% in the analysis of technical viscoses without excessive by-product carbon disulfide. Furthermore, to a certain extent the error is compensated for by calibration procedure. This error should be considered, however, if solutions containing large quantities of by-product carbon disulfide are to be analyzed.

The procedure of Dux and Phifer is not absolute, but must be calibrated in terms of another method. It has the advantage of being extremely rapid, however, taking only 20 min for one determination. The overall precision of the method is ±2.8%, which should be considered satisfactory for most industrial analyses.

The determination of secondary xanthate groups is best accomplished by the method described by Dunbrant and Samuelson.[24] A sample of viscose from which the by-product has been removed by the anion-exchange method described above is diluted with 0.1 M sodium hydroxide solution to 50 ml. A 5-ml aliquot of this is further diluted to 250 ml with 0.1 M sodium hydroxide at 50°C, and the solution is maintained at this temperature by a thermostat. The γ values are calculated by spectrophotometric analysis as before. The logarithm of the γ value is plotted as a function of time, and the straight-line portion of the curve extrapolated to zero time. The intercept on the ordinate corresponds to the number of primary xanthate groups.

SPECTROCHEMICAL CHARACTERIZATION OF REGENERATED CELLULOSE FILMS

The presence of xanthate and other sulfur compound groups in regenerated cellulose film can be identified by examining their ultraviolet absorption spectra. The analysis made by Kitamura and co-workers[25] has shown that both cellulose and alkyl xanthates have the same absorption spectra, characterized by two absorption bands, as viscose. The types of alkyl groups and univalent metals in the xanthate radical have no effect on the absorption spectra, but the second absorption band decreases in intensity when xanthates are hydrolyzed. In the case of alkyl xanthic acid, this band is scarcely recognized. Thiocarbonates and xanthates have second absorption bands in the near ultraviolet region, but both of them shift bathochromically. There is a distance of 15-20 $m\mu$ between the positions of the first absorption bands of xanthates and thiocarbonates. These facts confirm that the unexpected ultraviolet absorption is attributable to undecomposed cellulose xanthate remaining in regenerated cellulose films. Thiocarbonates have a third absorption band near 500 mμ, but xanthates do not have a corresponding band. This makes it possible to distinguish xanthate from thiocarbonate in the regenerated cellulose.

KINETICS AND MECHANISM OF THE FORMATION OF CELLULOSE XANTHATE

Various basic aspects of the xanthation reaction have been studied from time to time. The essential feature of this reaction is that with the action of carbon disulfide the crystallinity of cellulose progressively undergoes marked alteration. X-ray studies following the course of the reaction[26-28] have shown halo patterns with the reacted product and that the Bragg angle diminishes with xanthation. Lattice dimensions increase, and this increase continues with the ageing of the xanthate, presumably as a result of the production of sodium trithiocarbonate which expands the lattice. This widening of the lattice spacing is believed to be of critical importance during dissolution, since more open structures result in improved viscoses in terms of a more homogeneous viscosity. Xanthate that has been aged to a lower γ number develops and retains larger spacing and so dissolves better than a nonaged xanthate of a similar degree of substitution.

The xanthation reaction is believed to proceed gradually from the amorphous region into the micelle interior. The crystalline structure of such a reacted region is destroyed by distortion of the crystalline units and dissolution of the reacted chain by the alkaline medium. For 95% absorption of carbon disulfide by sodium cellulose, the lattice cell acquires a volume of 128.8 cubic A^o. The extent of lattice spreading depends upon the reaction temperature. The carbon disulfide finally distributes itself uniformly over all the cellulose. This is known as permutoid distribution, and the velocity of this permutite reaction is governed by the inter- and intramicellar diffusion velocity of carbon disulfide.[29]

A detailed kinetic study of the xanthation of simple carbohydrates and of cellulose has been described by Cherkasskaya and co-workers.[30] The reactions are carried out in sealed tubes. The reaction products are analyzed potentiometrically for the amount of xanthate, sulfide, and trithiocarbonate. The results indicate that the essential steps in the xanthation process are the same as those indicated previously (p. 411). For low degrees of esterfication the following expression is valid for the reaction velocity:

$$w = k_1 C_{CS_2}^{n}$$

The number n denotes the number of reacted OH groups and is found to be about 2 for sucrose, as well as for a degraded cellulose hydrate. The xanthation of the high-molecular-weight cellulose takes place under heterogeneous conditions and is therefore principally different. As shown by the previous investigators, this reaction has been confirmed to be permutoid and is considered pseudo-unimolecular, following the rate expression in which X is the

$$K_3 = \frac{2.303}{t} \log_{10} \frac{1}{1-X}$$

amount of combined carbon disulfide. This equation gives satisfactory values up to a degree of esterification of about 16%. Xanthation proceeds by two independent reactions: the reaction of carbon disulfide with carbohydrate and with excess alkali. In the first case the velocity of the xanthation passes through a maximum because the equilibrium is continuously disturbed by side reactions. An increase in alkali concentration decreases the reaction velocity in the xanthation of sucrose either because the OH groups are blocked by sodium hydroxide or because the equilibrium is shifted to the formation of sodium sulfide and thiocarbonate according to the following equations:

$$CS_2 + NaOH \longrightarrow S{=}C\begin{matrix}\diagup SNa \\ \diagdown OH\end{matrix} \qquad (9\text{-}7)$$

$$S{=}C\begin{matrix}\diagup SNa \\ \diagdown OH\end{matrix} \longrightarrow NaSH + COS \qquad (9\text{-}8)$$

$$NaSH + S{=}C\begin{matrix}\diagup SNa \\ \diagdown OH\end{matrix} \longrightarrow Na_2CS_3 + H_2O \qquad (9\text{-}9)$$

$$COS + 4NaOH \longrightarrow Na_2S + Na_2CO_3 + 2H_2O \qquad (9\text{-}10)$$

These side reactions have been shown to occur by an indopondont study by Grotjahn.[31]

The xanthation of alkali cellulose in a heterogeneous medium is much slower on the basis of equal amounts of reagent. A comparison of the velocity of reaction of carbon disulfide with sucrose, degraded cellulose hydrate, and alkali cellulose (denoted by A,B, and C, respectively, for the purpose of brevity) has been indicated in Tables 9-1, 9-1A, 9-1B and 9-1C.

TABLE 9-1

Kinetic Data on the Combination of Carbon Disulfide with Carbohydrate (A) Degraded Cellulose Hydrate (B) and Cellulose (C)

Xanthation of 0.575 mole A

Ratio A:CS_2:NaOH	Temperature, °C	Time, h	Moles of bound CS_2
1:1:3	40	1	0.31
		3	0.32
		6	0.22
		9	0.15
1:4:4	20	18	0.51
		24	0.71
		48	0.92
		96	0.53
		120	0.36
	40	168	0.23
		1	0.40
		3	0.75
		66	0.90
		9	0.58
1:12:12	20	24	0.40
		48	0.54
		96	0.90
		120	1.03
		168	1.20
		360	0.89
	40	1	0.22
		3	0.48
		6	0.93
		9	1.24
		18	0.52

TABLE 9-1A

Products Formed in the Xanthation of 0.575 mole A

Ratio A:CS_2:NaOH = 1:1:2

Temperature, °C	Time, h	No. of moles formed				
		Xanthate	Na_2S	Na_2CS_3	CS_2	NaOH
20	3	0.09	0.003	0.0015	0.492	1.067
	6	0.156	0.006	0.003	0.422	0.987
	9	0.210	0.009	0.0045	0.364	0.919
	18	0.245	0.012	-	0.265	0.791
30	0.5	0.085	0.0025	0.0005	0.498	1.076
	1	0.15	0.0051	0.0010	0.431	1.002
	2	0.24	0.0104	0.0021	0.336	0.892
	3	0.275	0.0158	0.0033	0.299	0.838
	5	0.302	0.0270	0.0059	0.260	0.769

Table 9-1A (con't)

40	0.25	0.135	0.006	0.0007	0.446	1.115
	0.50	0.230	0.012	0.0012	0.327	0.980
	1	0.325	0.024	0.0028	0.244	0.747
	2	0.350	0.059	0.0148	0.183	0.599
	4	0.320	0.1275	0.0218	0.168	0.402

The values of the first-order rate constant are 12.3×10^{-2}, 62.3×10^{-2}, and 281.0×10^{-2} at 20, 30°, and 40°C, respectively.

TABLE 9-1B

Products Formed in the Xanthation of 0.31 mole B
Ratio B:CS_2:NaOH = 1:3:8

Temperature, °C	Time, h	No. of moles formed				
		Xanthate	Na_2S	Na_2CS_3	CS_2	NaOH
15	1	0.036	0.005	0.001	0.890	2.447
	2	0.069	0.010	0.002	0.850	2.396
	3	0.099	0.015	0.003	0.806	2.347
	4	0.126	0.020	0.004	0.780	2.300
	5	0.150	0.025	0.005	0.761	2.266
25	0.5	0.040	0.006	0.001	0.878	2.430
	1	0.115	0,012	0.002	0.869	2.360
	1.5	0.143	0.018	0.003	0.785	2.295
	2	0.185	0.024	0.004	0.750	2.237
	3	0.185	0.036	0.006	0.0013	2.153

Table 9-1B (con't)

35	0.25	0.50	0.005	0.0013	0.877	2.431
	0.5	0.10	0.010	0.0025	0.818	2.325
	0.75	0.143	0.015	0.0038	0.771	2.300
	1	0.175	0.020	0.005	0.735	2.250
	2	0.210	0.037	0.010	0.679	2.143

The first-order rate constants are 4.2×10^{-2}, 11.0×10^{-2}, and 27.1×10^{-2} at 15°, 25°, and 35°C, respectively.

TABLE 9-1C

Xanthation of C

Ratio $C:CS_2:NaOH$ = 1:1:1.8

Temperature, °C	Time, h	Molar ratio product ($CS_2:C_6H_{10}O_5$)	First-order rate constant
20	1	0.091	11.0×10^{-2}
	2	0.182	
	3	0.287	
30	0.25	0.058	
	0.50	0.115	24.5×10^{-2}
	0.75	0.383	
	1	0.220	
	2	0.390	

Table 9-1C (con't)

40	3	0.510	
	0.25	0.144	64.3×10^{-2}
	0.50	0.283	
	0.75	0.383	

In general it is noted that there is an optimum cellulose:CS_2:NaOH ratio for the best yield of xanthated cellulose. An increasing free alkali concentration is not desirable in the xanthation reaction because it results in an increased formation rate for by-products. Transfer of carbon disulfide to the reactive sites of cellulose proceeds through water rather than sodium hydroxide. This is supported by the observation of relatively high activation energies,[32] which shows that the xanthation reaction occurs between the solid and the liquid phases.

The effect on the xanthation reaction of a double bond, a furan ring, a carboxyl group, and two -OH groups on the same carbon atom has been studied by Wronski.[33] The following rate expression is found to hold true for allyl alcohol and sodium glycolate:

$$\frac{k_1}{k} = \frac{1}{\xi} \ln \frac{a}{a-x}$$

where a = xanthate concentration, k_1 = xanthation constant, k = rate constant for the hydrolysis of carbon disulfide in sodium hydroxide solution, and ξ = concentration of by-products sodium sulfide and thiocarbonate. The reaction is more complicated in the case of furfuryl alcohol, monothiocarbonate being formed in addition to the other normal products and by-products.

Recently Andersson and co-workers[34] have studied the total rate of xanthation of low-molecular-weight cellulose in a homogeneous medium by spectrophotometric measurement using a stopped flow technique. By combining these results with determinations of the number of primary and secondary xanthate groups, the individual rate constants for the reactions of these groups can be calculated. It has been shown that in these systems the xanthate formation is a second-order reaction determined by the concentration of carbon disulfide and sodium alcoholate. The overall rate constant is given by the sum of two rate constants with respect to primary (γ_6) and secondary ($\gamma_{2,3}$) xanthate groups:

$$\left(\frac{d\gamma}{dt}\right)_x = \left(\frac{d\gamma_6}{dt}\right)_x + \left(\frac{d\gamma_{2,3}}{dt}\right)_x$$

For short reaction times the ratio between the rate constants is given by

$$\frac{k_{6x}}{2k_{2,3x}} = \frac{\gamma_6}{\gamma_{2,3}}$$

and is thus obtained directly from the relative number of xanthate groups determined by the kinetic decomposition method.

The values of rate constants of different alkali and carbon disulfide concentrations are given in Table 9-2.

It is seen that under otherwise unchanged conditions the cellulose concentration has no detectable influence on the rate constants defined above. This means that the rate of formation of xanthate groups is roughly proportional to the cellulose concentration. The rate constants are also virtually independent of the carbon disulfide concentration, but increase rapidly with an increase in the alkali concentration as a result of the increased formation of alcoholate groups. The ratio between the rate constants $k_{6x}:k_{2,3x}$ is not constant, but increases with an increased hydroxide concentration, indicating that the degree of dissociation of the primary hydroxyl groups increases

TABLE 9-2

Rate Constants of Xanthate Formation at 20°C at Varying Concentrations of Sodium Hydroxide, Cellulose, and Free Hydrogen Sulfide[a]

NaOH, M	Free CS_2, mM	Cellulose, %	k_{6x}, $1\ mol^{-1}\ h^{-1}$	$k_{2,3x'}$, $1\ mole^{-1}\ h^{-1}$
0.40	0.338	1.0	20.6	10.2
0.40	0.338	0.5	20.3	10.0
0.40	0.338	0.25	22.1	10.8
0.72	0.233	1.0	35.2	16.8
0.72	0.499	1.0	38.0	18.1
0.72	0.807	1.0	37.0	17.6
0.735	0.885	0.5	34.0	16.2
1.60	0.282	1.0	62.6	26.8
1.60	0.282	0.5	65.9	27.2
1.60	0.282	0.25	71.6	29.6
2.60	0.325	1.0	128.3	43.7
3.0	0.327	0.5	172.2	53.6
3.0	0.749	0.5	174.4	55.0

[a] After L. Andersson, O. Samuelson, and B. Törnell: Makromol. Chem., 119, 133-139 (1968).

more rapidly with an increased alkali concentration than does that of the secondary hydroxyl groups. It should be mentioned, however, that the rate constants decrease markedly with an increased degree of xanthation, which means that the rate of rexanthation in a technical viscose would be considerably lower.

SOME PHYSICOCHEMICAL ASPECTS OF RIPENING OF VISCOSE

As mentioned in the introduction, ripening of viscose precedes regeneration of ceululose and spinning. The chemical reactions and the changes in physical characteristics that occur during ripening are of importance in determining the optimum extent of ripening desired. The most important chemical reaction is the alkaline decomposition of xanthate, which was investigated by Klauditz.[35]

$$\underset{\substack{|\\ \text{O Cell.}}}{\text{S=C}} - \text{SNa} + 5\text{NaOH} \longrightarrow 2\text{Na}_2\text{S} + \text{Na}_2\text{CO}_3 + 2\text{H}_2\text{O} + \text{Cell.-OH} \qquad (9\text{-}11)$$

This is a slow reaction and proceeds over several days. The products of decomposition influence the physical characteristics of viscose. The reaction has been studied in detail by Dolby and Samuelson,[36] who have also detected the formation of sodium dithiocarbonate $CO(SNa)_2$ as the primary reaction product at a high alkali concentration (37%).

After dissolution of the cellulose xanthate in dilute sodium hydroxide, the viscosity of the viscose decreases for a certain length of time. It passes through a minimum and then increases until the viscose coagulates. The viscosity minimum increases with decreasing temperature, but the viscosity value at the minimum point is only slightly affected by temperature. The viscosity values and changes with time of ripening are also virtually independent of the degree of substitution (DS), thus indicating that the viscosity minimum is closely connected with the purely chemical reaction and not necessarily with micellar aggregates, as was believed earlier.

Lyselius and Samuelson[37] found that viscose treated with carbon disulfide and an addition of sodium hydroxide showed a viscosity minimum even after the viscose had passed the viscosity minimum before the carbon disulfide treatment. Furthermore, the viscosity minimum is less pronounced when an anion exchanger (which accelerates the dexanthation) is present in the viscose during ripening. The viscosity drop observed with carbon disulfide treatment can thus be attributed to rexanthation. This is not the only explanation, however, because a minimum is obtained even when dilute viscose is ripened in the presence of a large excess of ion exchanger, showing thereby that the viscosity of a dilute solution of cellulose xanthate does not necessarily increase with decreasing degree of substitution. In such a case the viscosity minimum is probably attributable to changes in the relative amounts of xanthate groups in the 2-,3-, and 6-positions.

The kinetics of dexanthation during the ripening of viscose has been studied by Dunbrant and Samuelson.[38] Dexanthation consists of two first-order reactions occurring simultaneously. The faster one is ascribed to the decomposition of secondary xanthate groups (present in the 2- and 3-positions)and the other to the decomposition of primary xanthate groups (present in the 6-position).

This is given by the expression:

$$\frac{d\gamma}{dt} = \frac{d\gamma_{2,3}}{dt} + \frac{d\gamma_{6}}{dt}$$

$$= -K_{2,3}\,\gamma_{2,3} - K_{6}\,\gamma_{6}$$

The reaction rate is affected by rexanthation processes occurring simultaneously. The latter, however, can be eliminated by ripening the viscose in contact with a strongly basic anion-exchange resin. The rate constants $K_{2,3}$ and K_6 are thus calculated from the rates of decomposition in the presence of resin, and these values are used to determine the γ_6 and $\gamma_{2,3}$ values for viscose ripened in the absence of resin; the method of analysis described above has been used. It has been found that the number of xanthate groups in the secondary positions decreases continuously during the ripening, whereas the number of primary xanthate groups passes through a maximum (see Fig 9-1). This result is likely to prove to be of great interest from a practical point of view.

From the difference between the actual rate of dexanthation found experimentally and the net rate of dexanthation obtained from Fig. 9-1, the

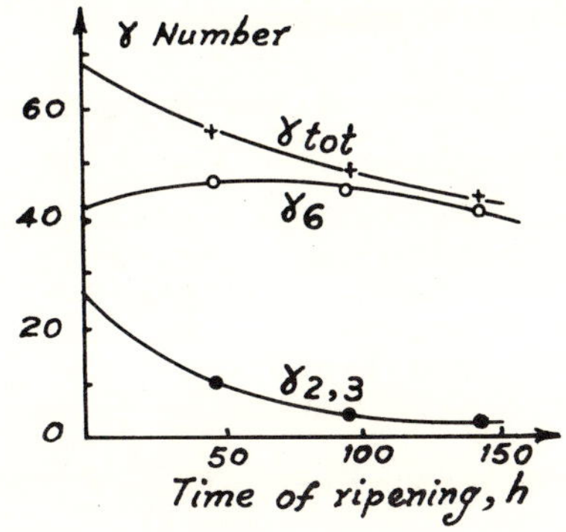

FIG. 9-1

γ Number as a function of the time of ripening for viscose ripened without resin. +, Total number; o, γ number of xanthate groups in the 6-position; o, γ number of xanthate groups in the 2 and 3 positions. [After S. Dunbrant and O. Samuelson: Tappi, 46, 520 (1963).]

rate of rexanthation in the 2- and 3-positions was calculated. It is found that dγ /dt varies very little with time (See Table 9-3).

The rate of dexanthation in the secondary position is high. However, there will always be some secondary xanthate groups present in the viscose.

The rate constants for dexanthation given by Dunbrant and Samuelson are: $K_6 = 3.5 \times 10^{-3}\ h^{-1}$ and $K_{2,3} = 46 \times 10^{-3}\ h^{-1}$. Slight variations are observed when viscose is treated with additional

TABLE 9-3

Calculation of the Rexanthation[a]

Time of ripening, h	Rexanthation in the 2- and 3-positions, % $\dfrac{\left(\frac{d\gamma}{dt}\ 2,3\right)_x}{\left(\frac{d\gamma}{dt}\right)_r}$	Rexanthation in the 6-position, % $\dfrac{\left(\frac{d\gamma}{dt}\ 6\right)_x}{\left(\frac{d\gamma}{dt}\right)_r}$	Total rexanthation, % $\dfrac{\left(\frac{d\gamma}{dt}\right)_x}{\left(\frac{d\gamma}{dt}\right)_r}$
0	48	27	75
47	37	29	66
114	33	24	57
144	31	20	51
340	-	-	45[b]

[a] After S. Dunbrant and O. Samuelson: Tappi, 46, 520 (1963).

[b] Calculated under the assumption that the presence of 2,3-xanthate can be neglected. The correct figure is higher (probably about 50) but cannot be determined unambiguously from available data.

quantities of carbon disulfide. With increasing sodium hydroxide concentrations in the concentration range used in technical ripening, both rate constants decrease. At 20°C, $K_{2,3} > K_6$ up to 6.2 moles free sodium hydroxide per liter, whereas at higher alkali concentrations the order is reversed; K_6 reaches a minimum at 4 moles and $K_{2,3}$ at 8 moles of sodium hydroxide per liter.

A detailed study of the rexanthation process that takes place during ripening has been recently carried out.[39-41] The rexanthation occurs via liberated carbon disulfide, as inferred from the observation that the concentration of free carbon disulfide in the viscose is much higher during an early stage of the ripening. At this stage there is a rapid loss of secondary xanthate groups and the rate of rexanthation in the 6-position is higher than the actual rate of dexanthation in this position. The rexanthation is promoted by subjecting the viscose sample to anion-exchange treatment before ripening, which takes place in the absence of resin. The anion-exchange treatment serves to exchange dithiocarbonate and sulfide ions formed according to Eq. (9-11) for hydroxyl ions. The rate of rexanthation decreases upon dilution of the viscose and with increasing sodium hydroxide concentration.

The distribution of xanthate groups with time of ripening is such that in a viscose with a normal alkali content the percentage of xanthate at C-6 approaches 100 after a fairly short time of ripening if the ripening is carried out under conditions such that rexanthation is eliminated. This is attributable to the much higher stability of the primary xanthate groups within this sodium hydroxide concentration range.

The rate of xanthate formation at the C-6 position decreases with decreasing reaction temperature. This is attributed to different activation energies for the rexanthation of the primary and secondary alcohol groups.[42-44]

The rate constants for the formation of primary and secondary xanthate groups under conditions close to those prevailing during viscose ripening have been determined by circulating a stream of gaseous carbon disulfide in nitrogen between a solution of carbon disulfide in paraffin oil and a diluted viscose.[41] The values for different alkali concentration ranges are shown in Table 9-4.

The rate constant values decrease markedly during the course of the reaction. This is partly

TABLE 9-4

Calculation of Rate Constants of Rexanthation[a]

Expt. NO.[b]	Time, h	6	2,3	Free CS_2, mmoles	Free NaOH, M	k_{6x} $1\ mole^{-1}\ h^{-1}$	$k_{2,3x}$ $1\ mole^{-1}\ h^{-1}$
1	0	37.3	21.6	3.91	0.75	15	7
	2.00	42.6	22.7	3.40	0.67	12	7
	4.00	45.5	22.7	3.79	0.60	9	6
	6.00	47.9	20.9	3.98	0.53	7	4
2	0	39.0	0.6	3.42	0.875	14	12
	2.00	43.7	10.8	3.49	0.82	13	8
	4.00	47.7	15.7	3.44	0.76	14	6
	6.00	52.3	16.7	3.78	0.64	11	5
	8.00	54.1	16.7	3.68	0.57	9	4

Table 9-4 (con't)

3	0	40.4	1.0	4.87	0.85	16	11
	2.00	48.0	16.6	5.04	0.72	14	7
	4.00	53.7	21.3	5.91	0.61	12	5
	6.00	58.9	23.2	5.94	0.52	9	4
	8.00	60.3	21.7	5.53	0.43	6	3

[a] After L. Andersson and O. Samuelson: Svensk Papperstid. 71, 727 (1968).

[b] Experiment 1: Freshly prepared viscose; 25 ml CS_2 in 200 g of paraffin oil.

Experiment 2: Ripened viscose treated with anion exchanger; 25 ml CS_2 in 200 g of paraffin oil.

Experiment 3: Ripened viscose treated with anion exchanger; 40 ml CS_2 in 170 ml of paraffin oil.

attributed to a decreased concentration of alcoholate ions. The degree of substitution also has a marked effect because of the hindrance to further xanthation created by the bulky xanthate ions. From the rate constants for the actual rate of dexanthation given previously (Table 9-3) and the rate constants for the formation of primary and secondary xanthate groups listed in Table 9-4, it is possible to make quantitative calculations of the changes in the number of primary and secondary xanthate groups. The choice of the proper value of k_x from Table 9-4 is difficult, but some simplification can be made by considering both k_{6x} and $k_{2,3x}$ as constants. The results of such calculations given by Andersson and Samuelson[41] confirm the rapid decomposition of the secondary xanthate and a less rapid decomposition of the primary xanthate, with the liberation of free carbon disulfide which in turn gives rise to a rexanthation in both positions. This rexanthation is of great interest from a practical point of view since it facilitates the dissolution of the xanthate and gives rise to a more uniform distribution of xanthate groups along the cellulose chains. This has a great influence upon the coagulation properties of the viscose.

DECOMPOSITION OF XANTHATE IN VISCOSE SPINNING

The reaction occurring in the process of spinning of the ripened viscose is essentially the reverse of xanthation. The cellulose xanthate is decomposed by a mineral acid, giving back cellulose and free carbon disulfide in accordance with Eq. (9-2). The presence of zinc sulfate in the system, however, complicates the mechanism of this reaction. Analytical studies[45] and physicochemical tests using radiotracers[46] showed that zinc cellulose xanthate does not form even as an intermediate product, but evidence for the formation of a complex has been advanced by the later investigators (see below).

The role of zinc ion in the decomposition of cellulose xanthate appears to be only to reduce the speed of decomposition. This aspect has been studied by Cichowski[47-50] in an extensive series of investigations on the influence of Zn^{2+} ion in decomposition of various xanthates in an acidic medium. The retarding effect is proportional to the concentration of Zn^{2+}. In an acidic medium xanthic acids are formed. Under the conditions prevailing during the fiber forming process, the reaction of Zn^{2+} with cellulose xanthate proceeds throughout the entire cross section of the spun fiber, whereby the cellulose xanthic acid is in a concentration-dependent equilibrium with Zn^{2+}. The retarding influence of Zn^{2+} is attributed to the binding of xanthic acid through zinc, probably resulting in a complex which is stable in acidic media.

The action of zinc in the dexanthation process has been studied by Törnell[51-53] by spectrophotometric analysis. The spectrum of the zinc-free solution is typical of a partly dissociated xanthic acid and has absorption maxima at 276 and 303 mμ, whereas the spectrum determined in the presence of zinc shows only one absorption maximum at 298 mμ. This suggests the formation of complexes between zinc ions and xanthate groups, and such chelate complexes exert an inhibitory effect on the dexanthation process. The strong complexing power of cellulose xanthate is explained by the high local concentration of xanthate groups in the macromolecules. The important influence of the local concentration of xanthate groups is demonstrated by the observation that the complexing power of a short-chain xanthate such as amyl xanthate is greatly increased by the addition of a nonionic surfactant. This effect arises because the xanthate accumulates in the surfactant micelles.

The strong complexing power of cellulose xanthate is demonstrated further by the observation that small amounts of zinc bring about extensive complex formation and a pronounced inhibition of the dexanthation reaction even in strongly acidic media (0.3 M H_2SO_4). Under these conditions the fraction of complexed xanthate groups can be determined from the spectra of the reaction mixtures. A comparison between the inhibitory effect of zinc and the extent of complex formation has shown that the retarding effect of zinc is even larger than that expected from a complete stabilization of the complexed xanthate groups. This may be related to the fact that in strongly acidic media complex formation results in a decrease in the number density of undissociated xanthic acid groups along the macromolecules.

It may be suggested that complexed xanthate groups are stable toward acid decomposition, and that the decomposition of cellulose xanthate occurs according to the following reaction scheme:[52]

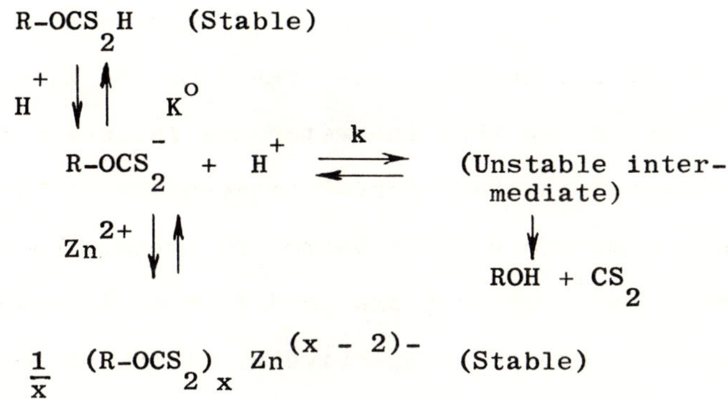

(9-12)

where K^o is the intrinsic dissociation constant of cellulose xanthic acid, k the second-order reaction rate constant, and x the number of xanthate groups should to each Zn^{2+} ion.

The ratio between zinc and xanthate is found to increase with increasing spinning distance, without a corresponding decrease in the specific rate of reaction.[51] It can therefore be inferred that the number of xanthate groups coordinated by each Zn^{2+} ion decreases during the decomposition.

Such a variation could be expected to occur as a result of the increase of the average distance between neighboring xanthate groups during the decomposition.

An estimate of the number of complexed xanthate groups per zinc ion, based on the kinetic data,[52] has shown that the retarding influence of zinc observed in the interval between the shortest spinning distance would require the assumption that each Zn^{2+} ion is bound to 4.6, 6.4, 4.7, and 4.9 xanthate groups, respectively, at the different zinc sulfate concentrations 0.1, 0.2, 0.4, and 0.8 g/liter. The number is 3.2 in the case of ripened viscose. As the normal coordination number of zinc is 4, it follows that at least at a high γ number the inhibitory effect of zinc is greater than expected from a complete stabilization of the complexed xanthate groups.

The stabilizing effect of zinc is found to increase with a decrease in the filament diameter and, up to a certain point, with an increase of the spinning distance. This is inferred from the fact that the fialment γ number reached at short spinning distances is practically unaffected by the presence of zinc at all but the highest concentration tested (0.8 g/liter $ZnSO_4$). These observations show that the rate of regeneration is determined in part by a diffusion process.

Zinc is found to accumulate in the filament phase. Under certain conditions the concentrations of zinc in the filament phase exceeds that in the spin bath solution by a factor of 100. It is strongly dependent upon the filament diameter and is higher with filaments of smaller diameter. The rate of uptake of zinc is rather low, and certainly much lower than the rate of acidification of filaments. It is noteworthy, however, that in the spinning of fine filaments in baths of not too low zinc contents the amount of zinc taken up by the filaments passed through maximum values at certain spinning distances. The presence of these maxima suggests an equilibrium distribution of zinc and the presence of zinc over the whole filament cross section.

SPINNING IN ZINC-FREE SPIN BATH SOLUTIONS

During spinning in zinc-free spin bath solutions, the dexanthation process follows first-order kinetics in most cases. The specific rate of decomposition is independent of the filament diameter, strongly indicating that dexanthation during spinning is largely independent of diffusion processes and that the filaments are completely acidified within a time interval much shorter than the half-life of the xanthate. This is shown by the fact that the sulfuric acid concentration is independent of filament diameter and, throughout the range of spinning distances studied, remains constant and very nearly the same as that of the spin bath solution. These observations suggest that an equilibrium distribution of electrolytes between the filament and spin bath phases is established during an early stage of regeneration process, i.e., before the majority of the xanthate groups are split off. It follows that under the conditions in a zinc-free bath dexanthation takes place in an almost homogeneous gel phase, with an electrolyte composition quite similar to that of the spin bath solution.

Addition of sodium sulfate to the spin bath solution brings about a slight decrease in the rate of dexanthation. The retarding influence of sodium sulfate upon the rate of dexanthation can be explained by a general salt effect, and by a decrease in the H^+ ion concentration because of the transformation of H^+ ions into HSO_4^- ions.

The degree of ripening of viscose strongly influences the rate constant of the dexanthation reaction.[52] In the presence of sulfuric acid (30 g/liter) and sodium sulfate (50 g/liter) a decrease in the viscose γ-number by ripening causes the rate coefficient to increase up to a certain stage and then to decrease again. In the absence of sodium sulfate, deviation from a first-order course of the decomposition is observed with both fresh viscose and viscose of the lowest γ-number. This shows that the reactivity of fresh viscose is initially low. The ability of viscose to deswell in the spin bath solution is strongly reduced by ripening.

Kinetics of the decomposition of cellulose xanthate has been studied by a number of investigators. Groebe et al.[54] have determined the effects of various parameters by following the reaction rate constant. This constant is found to be independent of the degree of ripeness and the initial degree of xanthation. It increases with a decreasing fiber diameter below 1 mm. At larger diameters the rate-determining factor is diffusion rather than cellulose xanthate decomposition. The relations between reaction rate constant (R) and temperature (T) are given by the expression

$$RT_2 = RT_1 \exp (A)/B\ (T_2 - T_1)$$

where B represents the gas constant and T the absolute temperature. The constant A is 17 ± 2 kcal $mole^{-1}$ for pure cellulose xanthate filaments. The relation between decomposition rate and pH are expressed by the equation

$$\log R = C\ pH + d$$

where C and d are constants depending on the decomposition of the xanthate. The regeneration of the cellulose xanthate at the pH of the spinning bath is the rate-determining step in filament formation.

Kinetics of the decomposition of cellulose xanthate determined by measuring ultraviolet absorption changes at 303 mμ has shown that the reaction proceeds in two steps[50]:

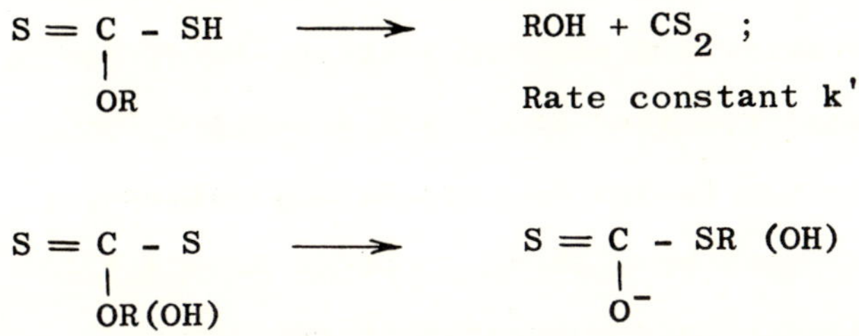

Rate constant k'

(9-13)

Rate constant k''

(9-14)

The pH of the solution affects k'' but not k'. The overall reaction rate constant is given by

$$k = k' \, (H^+) + k''$$

In an acidic medium the small difference in the decomposition rates of xanthic acid at primary or secondary substitution sites indicates only a weak contribution of the interaction between these sites to xanthate decomposition during viscose fiber formation.

The variation in the dexanthation rate constant with alkalinity has been explained by Elmgren[55] by postulating an equilibrium between xanthate and carbon disulfide plus alcoholate:

$$R - O^- + CS_2 \longrightarrow R - O - CS_2^- \qquad (9\text{-}15)$$

This equilibrium is disturbed by side reactions. The dissociation constants of the hydroxyls of the cellulose monomer are assumed to be 0.50 for the 6- (secondary) position and 1 and 4 for the two hydroxyls in the 2- and 3- (primary) positions. These values have been shown to be in accord with the dexanthation rate constants of cellulose xanthate as well as the activation energies for the overall dexanthation process. The following equilibrium expression relating the degree of substitution (DS) in 2-, 3-, and 6- positions has been established:

$$2/\ DS_{2,3} = 1 + \frac{k_6 \alpha_6 k'_{2,3}(1 - DS_6)}{k_{2,3} \alpha_2 \alpha_3 k_6 DS_6}$$

$$= 1 + \frac{B(1 - DS_6)}{DS_6}$$

In this expression $k_{2,3}$ and k_6 are the xanthation rate constants of the 2- and 3- and 6-positions, respectively. The product $\alpha_2 \alpha_3$ describes the probability that both the 2- and 3- positions are dissociated and α_6 the probability that the 6-position is dissociated. The k' terms refer to the dexanthation rate constants. The constant B has been shown to be approximately equal to $3\alpha_6 k'_{2,3}/\alpha_2 \alpha_3 k'_6$. The values of α are obtained from the experimental data of dexanthation at known hydroxyl ion concentrations. This expression can be used to predict the course of xanthation of alkali cellulose and of viscose ripening.

The formation of cellulose xanthic acid as an intermediate product in the decomposition of cellulose xanthate has been shown by Jost and Ludwig[56] by spectrophotometric analysis. Cellulose xanthic acid has an absorption maximum at 2740 A°. The value of the dissociation constant of cellulose xanthic acid is found to vary between 1.8 X and 13.2×10^{-2} at 0°C, depending upon the pH of the solution. This will be discussed in the next section.

CHEMISTRY OF CELLULOSE XANTHIC ACID

The kinetics of decomposition of cellulose xanthate in acidic media and the dissociation equilibrium of cellulose xanthic acid have been investigated by Törnell[57,58] by absorbance measurements. As mentioned before, cellulose xanthic acid shows an absorption maximum at 2740 A°

and xanthate ion at 3030 A^o. The absorbance measurements at these wavelengths of cellulose xanthate acidified with sulfuric acid showed that the degree of dissociation is lowest at the beginning of the reaction period and decreases with acidification. The acid strength of cellulose xanthic acid is influenced by at least two different effects. The first is attributable to the electrostatic interaction between polyions (cellulose xanthate ions) and the counterions. This effect is predominant in weakly acidic solutions and is determined by the charge density and structure of the polyions and the ionic composition of the solution. The charge density is expressed in terms of the product of degree of dissociation α and the γ -number. This is found to pass through a pronounced maximum (Fig. 9-2) during decomposition in solutions of ionic strength 0.1. This means that the number of charged xanthate groups increases initially in spite of the fact that the total number of xanthate groups decreases.

There is, however, no simple correlation between the degree of dissociation and the γ -number, a fact which is explained by suggesting that the

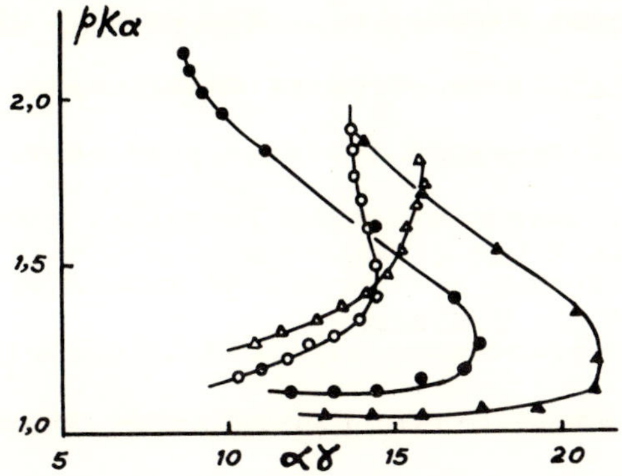

FIG. 9-2

Variation of pK_a and αγ during decomposition of viscose of types 1 (△ ▲) and III (O, ●) in 0.05 M HCl of 1.0 M (△, O) and 0.1 M (▲, ●) total chloride concentration at 25°C. [After B. Törnell: Svensk Papperstid. 70, (1967)]

concentration of the free xanthic acid groups influences the dissociation constant. This is the second effect on the dissociation constant and is referred to as the acidic group effect. At constant acidity the influence of this effect on pKa is nearly proportional to the number of uncharged xanthate groups per 100 anhydroglucose units. This effect decreases with increasing hydrogen ion concentration, but is still important within the acidity range of technical interest.

The formal dissociation constant of xanthate group in cellulose xanthate is expressed by Katchalsky's[59] formula

$$pKa = pH = \log (1 - \alpha) / \alpha$$

$$\text{or } pka = pK_0 + 0.4343 \alpha \gamma$$

where kK_o is an intrinsic dissociation constant and a constant dependent on ionic strength. The value of pKa when $\alpha\gamma$ and $(1-\alpha)\gamma$ both approach zero is difficult to determine accurately. It seems likely that it is in the range 1.0-1.2, suggesting that cellulose xanthic acid is stronger than ethyl xanthic acid (cf. Chapter 3).

A marked difference in the decomposition rate was found between fresh viscose and ripened vicose. The degree of dissociation also varies considerably during decomposition in acidic media (see Table 9-5).

No simple correlation exists between the rate coefficient and the γ-number. The influence of ripening on the rate of decomposition of cellulose xanthic acid cannot be explained on the basis of the decrease in the γ-number resulting from ripening, nor can it be related to the different

TABLE 9-5

Rate Coefficients and the Variation in the Degree of Dissociation during Decomposition in HCl-NaCl - Solutions of 1.0 M Total Chloride Concentration and in 0.3 M H_2SO_4 Containing Different Amounts of Na_2SO_4 at 25°C

$[H^+]$, M	$[Na_2SO_4]$, M	Unripened Viscose k, min^{-1}	Unripened Viscose α	Ripened Viscose k, min^{-1}	Ripened Viscose α
1.0		2.43	0.02-0.05	3.36	0.03-0.05
0.5		2.42	0.03-0.09	3.45	0.06-0.10
0.3		2.21	0.06-0115	3.03	0.10-0.14
	0	2.50	0.03-0.07	-	-
	0.1	2.51	0.07-0.17	-	-
	0.3	2.29	0.07-0.26	2.94	0.06-0.18
	0.5	2.29	0.09-0.23	-	-

relative amounts of primary and secondary xanthate groups in the fresh and ripened viscosed because within the acidity ranges tested (pH 0-3) the distribution of xanthate groups remains practically unchanged during decomposition.

It is not possible to express the acid decomposition of cellulose xanthate by a general rate expression. This is a consequence partly of an increase in the dissociation constant with a simultaneous decrease in the rate constant as the number density of uncharged xanthate groups decreases, and partly of other differences between the different samples such as the distribution of xanthate groups along the macromolecules. Considerable differences in behavior are observed between unripened and ripened viscoses.

It is interesting, however, to note that the variation in the second-order reaction rate constant (k) is of the same magnitude but opposite in sign to that of the dissociation constant K_o, so that the product kK_o remains comparatively constant. This is explained by assuming that the effective dielectric constant decreases with the increasing number density of uncharged xanthate groups.

The degree of dissociation of cellulose xanthic acid decreases with an increase in the γ-number. This is shown by the values of formal dissociation constants recorded in Table 9-6. As mentioned before, this decrease is explained by

TABLE 9-6

Formal Dissociation Constants of Cellulose Xanthic Acid

γ - Number	pK_a
105	2.07—2.24
89	2.04-2.01
68	1.86-1.88
49	1.24-1.70

the influence of the acidic group effect, which is caused by interactions of a long-range type.

The acidic group effect is ascribed to variations in the local dielectric constant inside the macromolecules. This was demonstrated by the immediate appearance of a turbidity when a solution of a highly substituted xanthate containing 0.06% cellulose was rapidly poured into an equal volume of 0.25 M hydrochloric acid. Presumably this reaction was caused by the transformation of dissociated xanthate groups into uncharged xanthic acid groups resulting from a change in the dielectric constant of the scattering particles.

Results also show that the xanthic acid group is much less polar than the carboxylic acid group. This nonpolar character can possibly be an important factor in determining the polymer-polymer interactions involved in viscose spinning. Under conditions in which the filaments become completely acidified before any appreciable decomposition of the xanthate has occurred, the fiber formation, the mechanical properties, and the swelling of the xanthic acid gel would be determined to a large extent by the number and properties of the xanthic acid groups attached to the macromolecules. This also explains the greater ease of formation of filaments during spinning from highly substituted viscoses than from viscoses of a lower degree of substitution.

The results recorded in Table 9-5 show that the addition of sodium sulfate causes the degree of dissociation of xanthic acid to increase markedly, while the effect on the reaction rate is small. However, at a lower sulfuric acid concentration (0.1 M) addition of sodium sulfate decreases the rate of reaction appreciably. In the presence of 0.5 M salt the time required for 50% conversion is more than twice as long as that in a salt-free solution. Sodium sulfate also has a pronounced influence on the degree of dissociation in this acid concentration range. The retarding action of sodium sulfate can be attributed to its buffering action, and this is confirmed by the fact that sodium chloride, which has negligible buffer action, does not significantly affect the rate of reaction in 0.05 M hydrochloric acid. The effect of sulfate is attributable to the fact that the second dissociation constant of sulfuric acid is comparable to that of the xanthic acid group. These results explain the retarding influence of sodium sulfate on the rate of dexanthation during spinning at a low concentration of sulfuric acid (see p. 446).

At low degrees of substitution the product kK_o, which can be regarded as a formal first-order reaction rate constant, depends upon the composition of the reaction medium. The variations observed at 25°C were in the range 6.2-9.4 min^{-1}. These values are similar to those found for ethyl xanthate at the same temperature (6.4-7.8 min^{-1}; cf. Chapter 3). With fresh viscoses, at high degrees of substitution, intramolecular interactions exert a stabilizing effect on the xanthate groups. This results in a pronounced decrease in the reaction rate and in an increase in the value of kK_o during the decomposition. This shows that in the case of highly substituted xanthate partial dexanthation by acid and by ageing yield products with different reactivities.

Differences in reactivity are also observed between cellulose xanthate from unripened and ripened viscoses with γ numbers of about 70 and 50, respectively. These differences can be ascribed to an increase in the spacings between neighboring xanthate groups during ripening. This observation is of practical importance in several recently developed spinning processes in which the viscose is spun without ripening and at high γ numbers. In these processes the stabilizing effect plays an important role in determining the rate at which cellulose is regenerated. In conventional viscose rayon processes, however, a similar influence would not be operative since the γ number is generally below 50.

The close-neighbor interactions resulting in a decreased rate of decomposition have been attributed to intramolecular association through hydrogen bonding between neighboring xanthate groups. This is supported by the observation that primary xanthate groups attached to adjacent anhydroglucose units can exert a stabilizing influence upon each other. The local xanthate concentration in cellulose xanthate molecules can reach rather high values. If the molecule is assumed to occupy a cylinder about 15 Å in diameter, the local xanthate concentration in a sample of γ number 70 has been estimated as 1.25 M, which is about 60 times that of a saturated aqueous solution of ethyl xanthic acid. This also strongly supports the possibility of intramolecular association mentioned above. Further confirmation of intramolecular association by hydrogen bonding has recently been given by Elmgren[60] in an independent study on intrinsic viscosities of cellulose derivatives.

OXIDATION REACTIONS OF CELLULOSE XANTHATE

The cellulose xanthate present in a viscose solution is susceptible to oxidation, producing dixanthogens similar to the simple dialkyl dixanthogens discussed in Chapter 5. In technical viscose, however, oxidation is not of practical significance because it is greatly hindered by the presence of other sulfur compounds such as sodium sulfide and thiocarbonate which are more susceptible to the action of oxygen and consume a large portion of it. Danilov and co-workers[61-64] have shown that a 100-g sample of plant viscose, relatively free from the above-mentioned sulfur compounds, absorbs 56.5 ml of oxygen, whereas addition of 3% sodium sulfite diminishes the oxygen uptake to 20.8 ml. Passage of oxygen into viscose results in a product of lower viscosity and has a lesser effect on the polymer content. Oxidation of cellulose xanthate for 24 h results in a 50-60% conversion to dixanthogen. Cellulose dixanthogen has also been prepared by treating an aqueous solution of sodium cellulose xanthate with diethyl dixanthogen. It gradually decomposes with evolution of carbon disulfide, and is decomposed by sulfuric acid, giving rise to a number of sulfur compounds such as hydrogen sulfide, carbonyl sulfide, carbon disulfide, and some elemental sulfur. Dissolved in sodium hydroxide it yields sodium cellulose xanthate as the primary product; thiocarbonate and sulfite (but no free sulfur) are also produced in small quantities. Similarly 20-25% aqueous ammonia reacts to form ammonium cellulose xanthate. With aniline, cellulose dixanthogen swells in 12-15 h and hydrogen sulfide is evolved. The solid product obtained is identified as cellulose xanthanilide, which is soluble in alkali and decomposed by hot dilute sulfuric acid with the formation of aniline, carbonyl sulfide, and hydrogen sulfide.

The presence of oxygen in viscose leads to the formation of polysulfide anions which react with carbon disulfide much more rapidly than do sulfide ions. Hence, the carbon disulfide in equilibrium with cellulose xanthate is removed more rapidly when polysulfides are present, and the ageing rate is increased. The formation of polysulfides can be prevented by the addition of sodium sulfite to viscose.

Danilov[63] has described the preparation of the thioanyhydride of cellulose xanthic acid. Vacuum-dried sodium cellulose xanthate is suspended in dry benzene and shaken with acetoacetyl chloride

(C_2H_5OCOCl) for 3-4 h; the resulting precipitate is ground in water and the residue washed with ether, yielding cellulose xanthic thioanhydride as a yellow solid. It is soluble in 20% ammonia and 7-10% sodium hydroxide. Action of aniline for 20-30 h gives cellulose xanthoanilide as an insoluble gray solid; its formation is accompanied by the evolution of hydrogen sulfide. The formation of a thioanhydride of cellulose xanthic acid has also been observed to occur from the cleavage of cellulose dixanthogens.[64]

PHYSICAL CHARACTERISTICS OF CELLULOSE XANTHATE IN SOLUTION

In aqueous solutions cellulose xanthate behaves as a polyelectrolyte, as shown[65] by the effect of the concentration of electrolytes on intrinsic viscosity. All electrolytes except strong alkalis show essentially the same reduction in intrinsic viscosity for solutions of equivalent ionic strengths. Hydroxyl ions produce a greater reduction, which may be the result of interaction with the hydroxyl groups of the cellulose in addition to the electrolyte effect on the xanthate groups. The change in intrinsic viscosity with concentration of electrolyte can be attributed to a change in the configuration of the molecule.

The internal structure of viscose is sometimes considered[66] a concentrated solution of polymer. It is greatly influenced by total alkali concentration, cellulose concentration, and the degree of polymerization.[67] The internal structure of the cellulose xanthate gel is influenced by that of viscose, however, the nature of the cellulose xanthate gel is determined by a system of coagulation.

The effect of different parameters on the cellulose structure has been studied in Wlodarski[68] by infrared spectroscopic measurements. A sudden decrease in the coefficients of intrinsic viscosity occurs after a xanthation period of 60-80 min, corresponding to a γ -value of 50-54 when the cellulose is regenerated with a strongly acidic coagulation bath (25 g H_2SO_4 and 7 g Na_2SO_4/ liter). With baths that cause slow coagulation and decomposition of viscose, celluloses with high intrinsic viscosity are obtained regardless of the degree of xanthation. There is a good possibility of obtaining hydrate celluloses with a high content of crystalline phase and well-formed crystallites.

The effect of after-ripening of viscose on intrinsic viscosity shows that there is an optimum condition for ripening that is suitable for the production of a fine-crystal structure characteristic of the cord fibers.

A detailed study of the polyelectrolyte behavior of a viscose solution has recently been made by Choudhury and co-workers[69-72] employing viscosity and light-scattering measurements. The values of intrinsic viscosity for the same concentration of cellulose in a water solution of cellulose xanthate are much higher than those in a 6% sodium hydroxide solution. This indicates a coiling of the polymer molecules in the presence of electrolytes as a result of the increasing attraction between charged ions. The viscosity characteristics are greatly influenced by the degree of ripening. From the effects of ripening on dissymmetry of light-scattering of viscose solutions, it is concluded that molecules are coiled up more during ripening. The high dissymmetry values are indicative of polydispersity and greater heterogeneity of the system.

The configurational and hydrodynamic parameters of sodium cellulose xanthate in dilute solution and the influence of the overall degree of substitution on these parameters have been evaluated[71] using viscosity, refractive index, and light-scattering data. The intrinsic viscosity values $[\eta]$ increase with an increasing degree of substitution. This indicates that in the chains of the cellulose derivatives some flexing or kinking is possible. When the degree of substitution is low, the interaction between solvent and solute molecules is small, polymer-polymer contacts are favored, and the polymer chains coil up, leading to a smaller hydrodynamic volume. However, as the number of xanthate groups in the chain increases, the interaction between a polymer segment and solvent molecules adjacent to it increases, and the polymer molecule tends to expand so as to reduce the frequency of contacts between pairs of polymer segments. The influence of the added steric restrictions to rotation about the chain bonds with increasing degree of substitution is presumably overshadowed by the increased polymer-solvent interactions.

The expansion factor is practically independent of molecular weight because of the inherent stiffness of the cellulose chain. The stiffness is characterized by the effective bond length.

The calculated values of this parameter suggest a high degree of coiling. The data obtained strongly support the view that cellulose xanthate in dilute solutions behaves as an extended, loosly coiled molecule, substantially stiffer than synthetic polymers but comparable to other cellulose derivatives in chain flexibility. This is in accord with the physicochemical properties of cellulose xanthate previously determined by ultracentrifugation,[73,74] electron microscopy,[75] and light scattering.[76]

Physicochemical characteristics of sodium amylose xanthate have been studied in dilute solutions by Pramanik and Choudhury.[72] From light-scattering measurements the weight-average molecular weight determined in dilute sodium chloride solutions varies with the concentration of the latter, decreasing with increasing concentration. This trend is explained by assuming that the amylose xanthate molecules in salt solutions show a preference for the water molecules and so the optical properties in the immediate neighborhood of the solute molecules are not representative of the medium as a whole.

TABLE 9-7

Molecular Parameters of Sodium Cellulose Xanthate in Dilute Sodium Hydroxide Solution[a,b]

DS	$[\eta]$	$\bar{M}_W$	$(S_2)^{\frac{1}{2}}$	α	$\phi \times 10^{-21}$
0.25	3.00	1.08×10^6	900.3	1.02	2.97
0.48	3.55	1.02×10^6	949.3	1.05	2.88
0.77	3.60	1.05×10^6	972.5	1.05	2.86
0.91	3.70	1.14×10^6	1025.0	1.03	2.69
0.26	2.15	3.12×10^5	539.1	1.01	2.92
0.42	2.20	2.63×10^5	551.6	1.02	2.35
0.69	2.25	2.78×10^5	561.3	1.01	2.41
0.90	2.45	3.33×10^5	598.4	1.02	2.59
0.29	1.30	5.26×10^4	155.8	1.01	1.28
0.47	1.30	5.71×10^4	153.8	1.03	1.44
0.79	1.40	5.56×10^4	162.1	1.03	1.24
0.94	1.55	5.88×10^4	174.1	1.03	1.18

[a] After B. Das and P.K. Choudhury: J. Polymer Sci. Pt. A-1, 5, 769 (1967).

[b] DS, Degree of polymerization; α, expansion factor; $[\eta]$, intrinsic viscosity; ϕ, hydrodynamic parameter; $\bar{M}_W$, average molecular weight; $(S_2)^{\frac{1}{2}}$, radius of gyration.

In 0.11 M sodium chloride, sodium amylose xanthate molecules appear to have a random-coil configuration. Two other configurational parameters, the effective bond length b, and the steric factor are found to be 6.24 and 1.020, respectively. It is concluded that the amylose chain in sodium amylose xanthate behaves as a typical flexible coil in dilute salt solution.

In emulsion-xanthated viscoses the cellulose xanthate is dispersed in the form of molecules in solutions. The crystallites are dissolved and the dilute solutions are molecularly dispersed even if the concentration of sodium hydroxide is extremely low.[77] At higher concentrations the solution may contain secondary aggregates. The molecular dispersion is always preceded by an intracrystalline swelling of the cellulose or partially xanthated cellulose. As a consequence, certain cellulose materials, e.g., cotton linters, need a higher concentration of sodium hydroxide than others (wood pulp and rayon) for complete dissolution because the intracrystalline swelling occurs at different concentrations.

REFERENCES

1. Cross and Bevan: British Patent 8700 (1892)

2. Samuelson, O.: Xanthation of cellulose by emulsified carbon disulfide. Svensk Kem. Tidskr., 61, 79-89 (1949).

3. Gartner, F. and O. Samuelson: Emulsion xanthation of cellulose. Svensk Papperstid., 53, 805-810 (1950); 54, 501-504, 588-590 (1951); Svensk Kem. Tidskr., 63, 114-120 (1951).

4. Samuelson, O., R.H. Anderson, and F. Gartner: Emulsion Xanthation of cellulose. Tappi, 37, 236-242 (1954).

5. Samuelson, O. and H. Jonen: Influence of variations in sodium hydroxide concentration during dissolution of cellulose xanthate. Svensk Papperstid., 55, 933-935 (1952).

6. Bartunek, R.: Potassium hydroxide cellulose and potassium cellulose xanthate. Papier, 6, 356-358 (1962).

7. Bartunek, R.: Xanthation of cellulose with various alkalis.Papier. 9, 254-262 (1955).

8. Schmiedeknecht, H. and W. Claus: The influence of wetting agents on the xanthation in emulsion. Faserforsch. Textiltech., 17(12), 571-576 (1966).

9. Tennant, H.R.: Regenerated cellulose as a (textile) finishing agent. Textile Ind., 131(7), 143-145 (1967).

10. Bridgeford, D.J.: Polymeric xanthate coated paper. U.S. Pat. 3,335,023, Aug. 8 (1967).

11. Allewelt, A.L. and W.R. Watt: New method for preparing cellulose thiourethanes. Ind. Eng. Chem., 49, 68-78 (1957).

12. Nepochatykh, V.I., Z.A. Rogovin, G.G. Finger, and E.M. Mogilevskii: Preparation of a copper xanthate fiber. Khim. Volokna, No. 2, 49-51 (1966).

13. Geiger, E. and B.J. Weiss. The preparation of highly xanthated cellulose xanthates. Helv. Chim. Acta, 36, 2009-2017 (1953).

14. Konkin, A.A. and Y.A. Rymashevskaya: Xanthation of cellulose. Zh. Prikl. Khim., 31, 459-465 (1958).

15. Mogilevskii, E.M., M.A. Ginzberg, and R.A. Khurgina: Temperature conditions for xanthation of cellulose. Khim. Volokna, No.2, 60-63 (1960).

16. Noguchi, T.: Influence of composition of alkali cellulose and xanthating conditions upon the degree of xanthation. J.Soc. Textile Cellulose Ind. Japan, 6, 153-155 (1950).

17. Yamada, K. and S. Mukoyama: Distribution of xanthate groups in cellulose xanthate. Kolloid Z., 163, 98-106 (1959).

18. Horio, M., R. Inamura, N. Komatsu, H. Sakata, and T. Kato: Investigation of the distribution of xanthate groups in various xanthates by means of paper chromatography. Sen-i Gakkaishi 19(2), 102-109 (1963).

19. Phillipp, B. and Chung-Hung Chu: Distribution of xanthate groups in the viscose process. Faserforsch. Textiltech., 16(5), 244-249, 495-501 (1965).

20. Schurz, J. and W. Zimmerl: Viscometric determinations of the degree of polymerization of organic xanthates. Papier, 12, 468-476 (1958).

21. Elmgren, H.: New methods for viscose analysis. Svensk Papperstid., 69, 225-231 (1966).

22. Samuelson, O. and F. Gartner: Utilization of ion exchangers in analytical Chemistry, XVIII. Acta Chem. Scand., 5, 596 (1951).

23. Dux, J.D. and L.H. Phifer: Determination of xanthate sulphur in viscose. Anal. Chem., 29, 1842-1845 (1957).

24. Dunbrant, S. and O. Samuelson: Determination of primary and secondary xanthate groups in cellulose xanthate. J. Appl. Polymer Sci., 9, 2488-2499 (1965).

25. Kitamura, R., M. Ichikawa, and M. Tanaka: Spectrochemical studies on regenerating cellulose films. Sen-i Gakkaishi, 18(8), 677-686 (1962).

26. Sakata, H. and N. Komatsu: The mechanism of cellulose reaction. Study of the mechanism of xanthation by X-ray analysis. Sen-i Gakkaishi, 18, 992-997 (1962).

27. Hess, K., H. Kiessig and W. Koblitz: The permutoid character of xanthation of fiber cellulose. Naturwissenschaften, 38, 433-434 (1951).

28. Noguchi, T.: Collapse of the micelle structure with xanthation. J. Soc. Textile Cellulose Ind. Japan, 6, 155-157 (1950).

29. Kuriyama, S. and E. Shiratsuchi: Mechanism of the xanthation of alkali cellulose fiber. J. Chem. Soc. Japan, Ind. Chem. Sect., 54, 62-64 (1951).

30. Cherkasskaya, P.M., A.B. Pakshver, and V.A. Kargin: The xanthation of carbohydrates and cellulose. Zh. Prikl. Khim., 26, 311-320 (1953).

31. Grotjahn, H.: The kinetics of the formation of cellulose xanthate Z. Elektrochem. 57, 305-317 (1953).

32. Pasteka, M., A. Weiss, A. Kramar, A. Piklerova, and A. Erbingerova: Kinetics of alkali cellulose xanthation. Faserforsch. Textiltech., 13, 450-453 (1963).

33. Wronski, M.: The kinetics of the xanthate reaction of allyl and furfuryl alcohol and glycolic acid. Lodz. Towarz. Nauk. Wydzial III, Acta Chim., No. 4, 47-54 (1960).

34. Andersson, L., O. Samuelson, and B. Tornell: Kinetics of the formation of cellulose xanthate in homogeneous medium. Makromol. Chem., 119, 133-139 (1968).

35. Klauditz, W.: Ripening of viscose. Papier, Fabr., 37, Tech.-Wiss. T1251 (1939).

36. Dolby, I. and O. Samuelson: Ripening of viscose. Formation of low molecular weight compounds. Svensk Papperstid., 69(9), 305-310 (1966).

37. Lyselius, A. and O. Samuelson: The ripening of viscose, I, II and III. Svensk Papperstid., 64, 145-151, 735-739, 815-818 (1961).

38. Dunbrant, S. and O. Samuelson: Ripening of viscose, IV. Distribution of xanthate groups and kinetics. Tappi, 46(9), 520-523 (1963).

39. Dolby, I., S. Dunbrant, and O. Samuelson: Ripening of viscose, V. and VI. Svensk Papperstid., 67(3), 110-112 (1964); 68(5), 136-140 (1965).

40. Andersson, L. and O. Samuelson: Influence of some parameters upon the rexanthation during viscose ripening. Cellulose Chem. Tech., II(3), 305-310 (1968).

41. Andersson, L. and O. Samuelson: Kinetics of the rexanthation. Svensk Papperstid., 71(20), 727-730 (1968).

42. Philipp, B. and Ke-Tsing Liu: Distribution of substituents in fibrous cellulose xanthate and viscose. Faserforsch. Textiltech., 10, 555-562 (1959).

43. Baer, H.J., H. Dautzenberg and B. Philipp: Kinetic Investigations of the transxanthation of cellulose xanthate. Faserforsch. Textiltech., 17(12), 551-558 (1966).

44. Philipp, B. and H. Dautzenberg: Xanthate group distribution and transxanthation in cellulose fiber xanthates and viscoses. Papier, 21(3), 118-124 (1967).

45. Groebe, A., H. Klare, and R. Maron: The problem of zinc cellulose xanthate formation during spinning of viscose. Faserforsch. Textiltech., 12(1), 29-30 (1961); 13, 1-9 (1962).

46. Groebe, A., H. Jost, K. Fryer, G. Hartmann, and G. Pfrepper: Intermediate formation of zinc cellulose xanthate during the spinning of regenerated cellulose filaments. Monatsber. Deut. Akad. Wiss. Berlin, 5(3), 170-173 (1963).

47. Cichowski, Z.: Decomposition of xanthate in viscose manufacture. Faserforsch. Textiltech., 13, 1393-396 (1962).

48. Cichowski, Z.: The decomposition of xanthates in the presence of zinc ions. Abhandl. Deut. Akad. Wiss. Berlin Kl. Chem. Geo. Biol., No.1, 187-192 (1963); No.3, 149-155 (1965).

49. Cichowski, Z.: Role of zinc ions in the decomposition of xanthates in acidic media. Polimery, 7(3), 92-97 (1962).

50. Cichowski, Z. and R. Grochowski: Kinetics of the decomposition of cellulose xanthate in a homoegenous medium. Polimery, 12(3), 102-106 (1967).

51. Törnell, B.: Decomposition of cellulose xanthate in viscose spinning process. Pure Appl. Chem., 14(3-4), 453-460 (1967).

52. Törnell, B.: Viscose spinning. V Svensk Papperstid., 70(15), 449-457 (1967) 70(16), 489-499 (1967).

53. Törnell, B.: Decomposition of cellulose xanthate during spinning of viscose solutions of low acid content. Svensk Papperstid., 70 (17), 1-7 (1967).

54. Groebe, A., H. Jost and H. Klare: Mechanism of viscose fiber formation. Faserforsch., Textiltech., 15(5), 193-201 (1964).

55. Elmgren, H.: The dexanthation rate constants. Svensk Papperstid., 70(3), 78-83 (1967).

56. Jost, H. and J. Ludwig: Cellulose xanthic acid, an intermediate product in the viscose spinning process. Faserforsch. Textiltech., 17(5), 194-201 (1966); 18(6), 274-278(1967).

57. Törnell, B.: Xanthate decomposition in acid media, Parts 3 and 4. Svensk Papperstid., 70, 268-277, 303-314 (1967).

58. Törnell, B.: Acid decomposition of highly substituted cellulose xanthate. J. Appl. Polymer Sci, 12, 2141-2152 (1968).

59. Katchalsky, A., N. Shavit, and H. Eisenberg: Dissociation of weak polymeric acids and bases. J. Polymer Sci., 13, 69 (1954).

60. Elmgren, H.: The position of intramolecular hydrogen bonds and their effect on the conformation of cellulose. J. Chim. Phys., 65, 206-211 (1968).

61. Danilov, S.N. and O.P. Koz'mina: Structure and decomposition of dixanthates. J. Appl. Chem. (USSR), 19, 1059-1078 (1946).

62. Danilov, S.N., N. Grad, and E.I. Geine: Oxidative reactions in alkali cellulose and viscose. Zh. Obshch. Khim., 19, 826-842 (1949).

63. Danilov, S.N., V.O. Klesman, and N.M. Grad: Thioanhydrides of xanthic acids and their transformations. J. Appl. Chem. (USSR), 22, 625-633 (1949)

64. Danilov, S.N., N.F. Gintse, and K.V. Levitskaya: Polysulphur compounds and composition of viscose studied with tracer atoms. Zh. Obsch. Khim., 28, 2948-2958 (1958).

65. Tait, C.W., R.J. Vetter, J.M. Swanson, and P. Debye: Physical characteristics of cellulose xanthate in solution. J. Polymer Sci., 7, 261-276 (1951).

66. Komatsu, N., O. Rizo, and S. Nonaka: The spinning mechanism of viscose. Sen-i Gakkaishi, 15, 532-539 (1959).

67. Elmgren, H.: The macromolecular properties of sodium cellulose xanthate in dilute solution. Arkiv Kem, 24(13), 237-282 (1965).

68. Wlodarski, G.: Modifications of cellulose structure during the viscose process. Abhandl. Deut. Wiss. Berlin, Kl. Chem. Geol. Biol., No.3, 125-131 (1963).

69. Ghosh, K.K. and P.K. Choudhury: Polyelectrolyte behaviour of viscose solution. Svensk Papperstid., 68, 72-75 (1965).

70. Ghosh, K.K. and P.K. Choudhury: Light scattering studies of viscose solution. Makromol. Chem,, 102, 217-221 (1967).

71. Das, B. and P.K. Choudhury: Molecular parameters of sodium cellulose xanthate in dilute solution. J. Polymer Sci., 5(A-1), 769-77 (1967).

72. Pramanik, A.G. and P.K. Choudhury: Physico-chemical and macromolecular properties of sodium amylose xanthate in dilute solution. J. Polymer Sci., 6(A-1), 1121-1134 (1968).

73. Claesson, S. and H.H. Bruun: Sedimentation velocity investigations on concentrated viscose solutions. Svensk Papperstid., 60, 336 (1957).

74. Brunn, H.H.: Viscose solutions. Sedimentation velocity investigations on concentrated solutions. Svensk Papperstid., 60, 567 (1957).

75. Ranby, B.G., H.W. Giertz, and E. Treiber: Electron microscopic examination of the viscose process. Svensk Papperstid., 59, 117, 205 (1956).

76. Onyon, P.F.: Light scattering measurements on viscose solutions. J. Polymer Sci., 37, 295 (1959).

77. Gartner, F. and O. Samuelson. The state of solution in emulsion xanthated viscoses. Sartryck Svensk Papperstid., 53, 1-3 (1950).

CHAPTER 10

MISCELLANEOUS APPLICATIONS OF XANTHATES AND DIXANTHOGENS

A number of minor uses of xanthates and dixanthogens have been described from time to time. Some of them are of potential industrial importance and may give rise to further uses in related fields of technology. A concise survey of such miscellaneous applications will be made in this chapter.

APPLICATIONS IN RUBBER INDUSTRY

Zinc butyl xanthate has been used as an accelerator in the vulcanization of rubber.[1-4] In one process a solution of solvent naphtha containing 5% each of zinc butyl xanthate and piperidine pentamethylene dithiocarbamate is applied to a rubber sheet. After coating with the solution, the sheet is allowed to stand for 30 min and then vulcanized by heating for 1 h at 50°C to obtain an elastic vulcanized sheet.

10. MISCELLANEOUS APPLICATIONS OF XANTHATES

Another use of xanthates has been found in the protection of rubber against atmospheric gases, especially oxygen and ozone.[5,6] For this purpose nickel isopropyl xanthate in a concentration of 3%, based on the weight of rubber, is added as a blending agent to the rubber latex, and the films prepared from this latex are elongated in one dimension. Some of the long-chain dixanthogens have been used in a similar manner; the examples reported are ethyl butyl dixanthogen and dicyclohexyl dixanthogen.[7] About 1.5% , based on the weight of rubber, is added to SBR latex and stirred for 2 h. The compounded rubber specimens are more stable toward outdoor weathering and exposure to light than those stabilized with conventional stabilizing agents.

Dixanthogens have been used for regulation of the molecular weight of styrene butadiene rubber. The regulating efficiency of a series of dixanthogens has been studied by Vaclavek.[8] He has shown that the regulating efficiency of dixanthogens decreases with increasing chain length. The efficiency of any dixanthogen is lower for cold polymerization than for reaction at elevated temperatures. Regulation is effected through the reagent's capacity to promote reactions that interrupt the growth of the polymeric chain in the condensation polymerization process. The action depends upon its passage through the phase boundary of monomer droplets, diffusion through the water phase, and through the phase boundary of monomer-polymer particles, and chain transfer and growth reactions which take place in the particles. The slowest step of the whole process is diffusion through the water phase, the mechanism of which has not been explained in terms of the chemical reactions involved.

XANTHATES AS CATALYSTS IN THE POLYMERIZATION OF OLEFINS

Polymerization of olefin oxides and of olefin sulfides to high-molecular-weight polymers is catalyzed by certain heavy metal xanthates, zinc isopropyl xanthate being the best example.[9,10] A seeded catalyst is prepared in a nitrogen atmosphere by injecting 12.0 ml of propylene oxide into a $0^{\circ}C$ solution of 16.0 g of zinc butyl xanthate in 160 ml of dry benzene. The molar charge ratio of propylene oxide to zinc butyl xanthate in the seeded catalyst is 3.9. The solution is allowed to age for 2 days, during which it remains homogeneous.

In place of zinc isopropyl xanthate, diisopropyl dixanthogen has also been used as a catalyst in the polymerization of rubber.[11] The mechanism of these catalytic activities is still obscure. It would be of interest to investigate the role of an isopropyl group linked with a xanthate group and to determine why this particular combination is the most effective catalyst.

INSECTICIDAL AND PESTICIDAL USES

For many years xanthates and xanthic acid esters have been used as weed killers. Mixtures of sodium or potassium and ammonium xanthates are mixed with managnese, zinc, or iron salts. The resulting mixtures, used directly or after incorporation of liquid lime, are effective pesticides.

The methyl ester of propyl xanthic acid has been described as an effective insecticide. At 12.5 g/cm^2 it gives 100% control of houseflies in 18 h, at 100 g/cm^2 it gives 100% control of

Aphis fabae in 6 h, and at 0.2 g/cm^2 it gives 100% control of the confused flour beetle in 10 days. Other xanthic esters have also been tried as insecticides with varying degrees of success. Methyl esters of thioxanthic acids are used as highly effective insecticides against flies, wheat weevils, lice, and rice weevils and as parasiticides to kill alphids on bean plants. They are prepared by mixing a thio alcohol, potassium hydroxide, and carbon disulfide in stoichiometric amounts to form xanthate. Dimethyl sulfate is then added, the temperature maintained at 60°-$70^\circ C$ for 1 h, the solution cooled, and the product extracted with benzene.

USE OF XANTHATES IN THE INITIAL GROWTH OF PLANTS[15]

The germination and initial growth of corn, pea, wheat, and sugar beet can be greatly increased by treating water-soaked seeds with a 0.001% solution of a copper or iron salt of cellulose xanthic acid. Seeds of peas and corn treated with ferrous cellulose xanthate produce the longest roots and stems. The cellulose xanthate of copper (II) or iron (II) is synthesized by the reaction of technical sodium cellulose xanthate and 10% solutions of cupric sulfate or ferrous sulfate.

LUBRICATING OIL ADDITIVES

Dixanthogens and xanthic acid esters are used as additives in lubricating oil.[16-19] They possess good antiseize properties. The use of isopropyl, butyl, amyl, and nonyl dixanthogens has been described.[16] The antisieze properties decrease with the length of the hydrocarbon chain. Dibutyl dixanthogen has been found to be particularly useful as an extreme-pressure lubricant additive; it is characterized by thermal stability, as the decomposition temperature lies above 200°C. Lubricating properties are improved by the addition of xanthic esters to dixanthogens. In particular, decyl esters of nonyl and amyl xanthic acids, $C_9H_{19}OCS_2C_{10}H_{21}$ and $C_5H_{11}OCS_2C_{10}H_{21}$, are used for their stabilization and antiwear efficiency.

FIREPROOFING AGENT

Dibutyl dixanthogen is used as a constituent of a fire proofing agent for self-extinguishing resins.[20]

CORROSION INHIBITORS[21]

N-alkyl pyridinium alkyl xanthates with the general formula

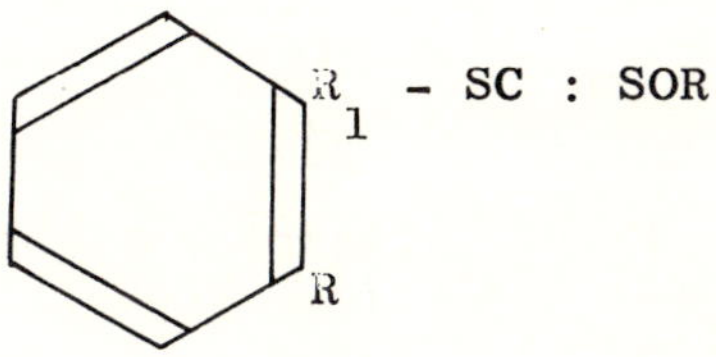

have been found to be efficient corrosion inhibitors for steel in acid solutions. They are prepared by the reaction of N-alkyl pyridinium bromide (10^{-2} mole) with equimolar amounts of sodium or potassium alkyl xanthates in 15-17 ml of boiling acetone or 12 ml of methanol. After refluxing for 5 min the inorganic salts are filtered off and washed with hot solvent. The products crystallize from the combined filtrates on cooling and are recrystallized from acetone-diethyl ketone or methanol-diethyl ketone mixtures. If water is used as a solvent for carrying out the reaction, the oily product is isolated by extraction with chloroform and purified by crystallization from an acetone-diethyl ketone mixture.

10. MISCELLANEOUS APPLICATIONS OF XANTHATES

REFERENCES

1. Ginzburg, L. Ya.: Cement SVKG for shoes. Legkaya Prom. 10(4), 16-17 (1950); Chem. Abstr., 47, 3600d (1953).

2. Shimosato, J.: Vulcanization of rubber. Japanese Pat. 2990, May 29 (1954).

3. Howorth, J. and W.R. Pryer: Staining problems associated with rubber chemicals. Trans. Inst. Rubber Ind., 25, 265-286 (1949).

4. Groux, G. and S. Lemarchand: A study of ageing of manufactured products from trypsinated latex. Rev. Gen. Coautchouc, 28,867-871 (1951); Chem. Abst., 46, 2327C (1952).

5. Karmitz, P.: Rubber protection against atmospheric agents. Rev. Gen. Caoutchouc, 35, 1243-1251 (1958).

6. Znamenski, N.N., L.S. Fomina, and V.V. Chernaya: The ozone and light stability of films of latex. Kauchuk i Rezina, 21(6), 6-8 (1962).

7. Hufenreuter, W. and W. Gruneberg: Stabilizers for synthetic rubbers. East German Pat. 50,305, September 20 (1966).

8. Vaclavek, V.: Regulation of molecular weight of styrene butadiene rubber. J. Appl. Polymer Sci., 11(10), 1881-1902 (1967).

9. Joginder, L.: Polymerization of olefin oxides and sulfides with xanthates. J. Polymer Sci., 5(B), 793-798 (1967).

10. Joginder, L: Polymerization catalysts for olefin oxides and sulphides. French Pat. 1,446,901 (cl.c08g), July 22 (1966).

11. May, D.R. and I.M. Kolthoff: Determination of disappearance of bis isopropyl xanthogen. J. Polymer Sci., 4, 735-743 (1949); Chem. Abstr., 44, 4708 (1950)

12. Nash, L.H.: Pesticidal ammonium alkyl xanthates. U.S. Pat. 2,900,293, Aug. 18(1959).

13. Wildgrube, W. and M.Born: Insecticides. East German Pat. 50,152, Oct. 25 (1966).

14. VED Farbenfabrik, W.: Pesticidal xanthic acid ester. British Pat. 1,077,277, July 26 (1966). French Pat. 1,450,866, Aug. 26 (1966).

15. Lisoval, A.P. and L.I. Rastorgueva: Influence of polyvalent metal salts of cellulose xanthic acid on the initial growth of plants. Visn. Sil's'kogospodar. Nauki, 6(1), 111-113 (1963); Chem. Abstr., 59, 1070g (1963).

16. Khalikov, R. Kh. and I.E. Vinogradova: Reaction of metals with organic sulfur compounds. Khim. Tekhnol. Topliv i Masel, 9(2), 63-67 (1964).

17. Harle, O.L.: Antioxidants for lubricating oils. U.S. Pat. 2,681,316, June 15 (1954).

18. Rabl. V., J. Cizek, and J. Mostecky: Thermal stability of extreme pressure (lubricant) additives. Sb. Vysoke Skoly Chem. Technol. Praze, addil Fac. Technol. Paliv Vody, 9, 21-30 (1966).

19. Fal'kovskaya, A.C., M.Sh. Oberfel'd, G.M. Kheifets, and L.B. Rapoport: Tr. Vses. Nauch. Tekh.Soveshch. Prisadkam Miner. Maslam, 2nd, 1966, 67-73 (1966).

20. Eichhorn, J.: Polysulphide fire proofing agents for self-extinguishing resins. U.S. Pat. 3,284,544, Nov. 8 (1966).

21. Palecek, U., A. Marhoul, J. Nemeova, and V. Sourek: N. alkyl pyridinium alkyl xanthates. Chem. Prumysl, 16(10), 588-590 (1966).

AUTHOR INDEX

Numbers in parentheses are reference numbers and indicate that an author's work is referred to although his name is not cited in the text. Numbers that are underlined show the page on which the complete reference is listed.

A

B

SUBJECT INDEX

A

B

C

R

S

T